CARVING OUR DESTINY

SCIENTIFIC RESEARCH FACES A NEW MILLENNIUM

Susan M. Fitzpatrick and John T. Bruer, editors

JOSEPH HENRY PRESS
Washington, D.C.

Joseph Henry Press • **2101 Constitution Avenue, N.W.** • **Washington, D.C. 20418**

The Joseph Henry Press, an imprint of the National Academy of Press, was created with the goal of making books on science, technology, and health more widely available to professionals and the public. Joseph Henry was one of the founders of the National Academy of Sciences and a leader of early American science.

Any opinions, findings, conclusions, or recommendations expressed in this volume are those of the author and do not necessarily reflect the views of the National Academy of Sciences or its affiliated institutions.

Library of Congress Cataloging-in Publication Data

Carving our destiny: scientific research faces a new millennium / Susan M. Fitzpatrick and John T. Bruer, editors.
    p. cm.
    Includes bibliographical references.
    ISBN 0-309-06848-7
    1. Science. I. Fitzpatrick, Susan M., 1956– II. Bruer, John T., 1949–
III. Title.
    Q171 .C48 2000
    500—dc21   00-011403

Cover illustration: *Man Carving Out His Own Destiny*, sculpture by Albin Polasek.

Printed in the United States in America

Contents

iii

Preface

In December 1998, the James S. McDonnell Foundation announced the 10 recipients of its James S. McDonnell Centennial Fellowships. The fellowships commemorate the 100th anniversary of James S. McDonnell as well as the 50th anniversary of his establishing the foundation that today carries his name. The fellowships awarded $1 million for each fellow's research program. A key component of the Centennial Fellows competition was each candidate's essay, intended for the general, educated reader, describing the planned research and explaining its significance. The winning essays are presented in this volume, updated and revised for publication following the April 9–10, 1999, Centennial Fellowship Symposium held at the National Academy of Sciences.

Although there were five fellowship categories, in this volume the essays are divided into four sections, reflecting how the fellows presented their work at the Centennial Fellows Symposium. As it turned out, the Centennial Fellowships in History and Philosophy of Science had an interest in genetics and cognition. Neurophilosopher Kathleen Akins' work on color vision is placed, as was her presentation at the symposium, in the same section as the essays by the Human Cognition Centennial Fellows. Keith Wailoo, the Centennial Fellow in History of Science, studies diverse issues in medical science, including genetic disorders, and his presentation at the symposium and his essay in this book, accompany those of the Human Genetics Centennial Fellows.

From the earliest planning stages of the Centennial Fellowship Program, the foundation intended to publish a volume containing the 10 winning essays. After the conclusion of the symposium, many attendees encouraged us to include the insightful and thought-provoking summary remarks made by the four senior commentators, Patricia Churchland, David Wilkinson, David Schlessinger, and Murray Gell-Mann. There were also many requests to include the tribute to the late James S. McDonnell that was prepared and presented at the symposium by his son, John F. McDonnell. The details of "Mr. Mac's" life and career provide a compelling rationale as to why the Centennial Fellowship Program chose to commemorate the past by investing in the future research efforts of 10 extraordinarily bright and articulate young scientists. Born in the last year of the nineteenth century, Mr. Mac's ability to envision the future would make him a leader in the twentieth century's most remarkable technological achievements—air travel and space flight.

The Centennial Fellowship Program and the Centennial Fellowship Symposium required four years of planning and implementation. It owes its success to the dedicated efforts of many individuals. First, we thank those scientists who responded to the foundation's call for applications. The academic community's response to the Centennial Fellowship Program exceeded our expectations. It was gratifying to have such abundant evidence for our belief that there are many young investigators who are both outstanding scientists and gifted communicators. We are very grateful for the efforts of the members of the advisory panels, listed below, who thoughtfully read, reviewed, discussed, and agonized over the applications. They had a most difficult job, made all the harder by their unwavering dedication to the foundation's goals.

HUMAN COGNITION

Endel Tulving, Rotman Research Institute of Baycrest Centre
Gordon Bower, Stanford University
Patricia Smith Churchland, University of California, San Diego
Max Coltheart, Macquarie University
Riitti Hari, Helsinki University of Technology
George Miller, Princeton University
Steve Petersen, Washington University School of Medicine
Steven Pinker, Massachusetts Institute of Technology
Michael Posner, University of Oregon
Larry Weiskrantz, University of Oxford

ASTROPHYSICS AND COSMOLOGY

David Schramm, University of Chicago (deceased)
Roger Blandford, California Institute of Technology
Marc Davis, University of California, Berkeley

Margaret Geller, Harvard-Smithsonian Center for Astrophysics
Richard McCray, University of Colorado
Sir Martin Rees, Cambridge University
Katsuhiko Sato, The University of Tokyo
Rashid Sunyaev, Russian Academy of Sciences and Max Planck Institute
    for Astrophysics
Edward van den Heuvel, University of Amsterdam

HUMAN GENETICS

David Schlessinger, National Institute on Aging
Sir David Weatherall, University of Oxford
Jorge Allende, Universidad de Chile
Kay Davies, University of Oxford
Mary-Claire King, University of Washington
Jean-Louis Mandel, Institute of Genetic, Molecular and Cellular Biology
Ulf Petterson, University of Uppsala Biomedical Center
Janet Rowley, University of Chicago Medical Center
Yoshiyuki Sakaki, University of Tokyo
Huda Zoghbi, Baylor College of Medicine

GLOBAL AND COMPLEX SYSTEMS

Peter Raven, Missouri Botanical Garden
Partha Dasgupta, University of Cambridge
Gilberto Gallopin, Stockholm Environment Institute
Murray Gell-Mann, Sante Fe Institute
John Holland, University of Michigan
Crawford Holling, University of Florida
Robert Kates, College of the Atlantic
Jane Lubchenco, Oregon State University
Jessica Tuchman Mathews, Carnegie Endowment for International Peace
Berrien Moore III, University of New Hampshire
F. Sherwood Rowland, University of California, Irvine
Alvaro F. Umana, INCAE

HISTORY AND PHILOSOPHY OF SCIENCE

Daniel Kevles, California Institute of Technology
Phillip Kitcher, University of California, San Diego
Nancy Cartwright, London School of Economics
Loren Graham, Massachusetts Institute of Technology
Clark Glymour, Carnegie Mellon University

John Heilbron, University of California, Berkeley
David Malament, University of Chicago
Mary Jo Nye, Oregon State University
Dominique Pestre, Centre de Recherche en Histore des Sciences et des Techniques
Elliott Sober, University of Wisconsin, Madison

The decision to administer the Centennial Fellowship Program using only the foundations' existing staff meant additional work for already busy people. Kathy Leonard, Michael Lee, and Cheryl Washington cheerfully and ably shouldered the many tasks required for the foundation to run an efficient and successful international program. Alene Roth deserves special recognition for her untiring efforts on behalf of the program and for her professionalism in handling the myriad logistical details of the advisory board meetings and the Centennial Fellowship Symposium. Much gratitude is also owed Program Officer Melinda Bier for her invaluable assistance on every aspect of the program.

The Centennial Fellowship Symposium was cosponsored by the National Academy of Sciences, and we are grateful to the many individuals there who contributed to its success. We are grateful for the enthusiastic support provided by Academy President Bruce Alberts, Vice President Jack Halpern, and Executive Director Ken Fulton. Donna Gerardi, director of the Academy's Office of Public Understanding of Science, went far beyond what the normal duties and responsibilities of her office required to help us with the Centennial Fellowship Program. From the earliest planning stages Donna "got" what the foundation was trying to accomplish and provided invaluable advice and guidance to the program. The success of the symposium is largely due to the help Donna and her staff provided.

The foundation also thanks Allison Ward and Dale Didion for assistance with public outreach. We are grateful for the interest and support of Stephen Mautner, of the Joseph Henry Press, for his enthusiastic assistance with this book.

S. M. Fitzpatrick
J. T. Bruer

Introduction

On June 10, 1995, the James S. McDonnell Foundation Board of Directors first discussed what the foundation might do to commemorate the centennial of James S. McDonnell's birth, in 1999, and the 50th anniversary of his establishing the foundation, in 2000. In that initial discussion, the board decided that any planned centennial program would have to look, and contribute, to the future, not merely commemorate the past.

The specific idea for a Centennial Fellowship Program arose a few months later during a visit by John Bruer, the foundation's president, and Susan Fitzpatrick, the program director, to Oxford University. They had spent the morning meeting with several of the outstanding young scientists that the foundation was supporting at Oxford. During a cab ride from St. John's College to the River Meade Rehabilitation Hospital, an idea began to take shape: An international competition to provide generous and flexible research support to young scientists would be an ideal, future-directed centennial activity. The foundation could award centennial fellowships in research areas that had been of special interest to Mr. McDonnell. In accordance with the board's wishes, and in recognition of Mr. McDonnell's reputation for forward-thinking, the fellows would be chosen not only on the basis of their scientific accomplishments, but also on the basis of the potential impact of their work for twenty-first century science. In addition, it was decided that the fellows would be articulate young people, dedicated to communicating both the importance of their research and the excitement of science to a broad public. To emphasize the importance of this criterion, the application process would not ask for a traditional research proposal; rather, it would require an essay that is accessible to a broad, science-interested audience that explained the fellows' research and its importance for society's future. The fellows would be selected then, in part, for their scientific promise and in part for their willingness to articulate how science and technology are central to addressing and solving social and human problems, now and in the future.

After some refinement of these ideas, the Board of Directors approved the Centennial Fellowship Program in fall 1995. Four years of intense preparation and deliberation followed. Finally, on April 9–10, 1999, on the centennial of James S. McDonnell's birth, in the gracious halls of the National Academy of Sciences, the 10 James S. McDonnell Centennial Fellows presented their work in a series of public symposia.

Selecting 10 fellows from among the hundreds that applied was a demanding task. It required months of reading and weeks of discussion among foundation staff and its advisors. However, our advisors took on the task in good humor and with an optimistic outlook. Most importantly, however, throughout the entire process the advisors took the time and effort to educate us about their various fields. They helped us to see the important and interesting problems that a private foundation could address. This learning process provided by our advisors reinforced the belief at the McDonnell Foundation that there are numerous opportunities where targeted and timely private foundation funding can make meaningful contributions to the advancement of science. Thus, one message we hope the Centennial Fellowship Program will articulate is that private foundations have a significant role in funding science and scholarship.

Organized American philanthropy is also celebrating its centennial. During its first 50 years, American foundations were prominent in the support of science, as is evident to anyone familiar with the histories of, for example, the Rockefeller and the Carnegie foundations. However, following World War II, with greater federal involvement in funding the national research effort, foundations began to look for other ways to invest their resources, resources that were now dwarfed by the federal commitment. Since the 1960s, many foundations have pursued policy-directed work and direct-action programs, with beneficial and commendable results. But, in a time of expanding wealth and of the imminent intergenerational transfer of that wealth with its enormous philanthropic potential, we should not forget that the original vocation of American-organized philanthropy was to invest in generating new knowledge that could be used to solve social problems. This vocation was at the core of James S. McDonnell's philanthropic vision. It was captured best in a remark he made in his 1963 commencement address to graduates of Washington University: "As man recognizes his responsibilities and moves ahead to carve his own destiny, there will be many problems to solve."

We have used this remark as the guiding theme for the Centennial Fellowship Program. James S. McDonnell believed that science and scholarship gave mankind the power to shape the future and the obligation to shape that future in a responsible manner, to the benefit of all. The fellows' research, as described in the chapters presented in this volume, exemplifies his ideal.

Tribute to James S. McDonnell

John F. McDonnell

James S. McDonnell often referred to himself as a plodder (namely, a person who works slowly, steadily, and unimaginatively). He did everything in life with meticulous attention to detail, to the point that it could be excruciatingly, maddeningly exasperating to those around him; but he also inspired those same people with a sense of important mission and high purpose. He had a unique ability to synthesize the details and see the whole picture and future evolution with an uncanny accuracy. And he had an insatiable, searching curiosity that enveloped everything about human existence from genetics and evolution through the interrelationship of the mind and brain, from the complexity and interconnectedness of everything on our spaceship Earth (his term) through the vastness of the cosmos, encompassing the fundamental philosophical questions of science and human existence. In this tribute I hope to provide some feel for the breadth and scope of the man who made the McDonnell Centennial Fellowships possible and how enthusiastic and knowledgeable he was about the fields of discovery that are presented in this volume: human cognition, astrophysics and cosmology, human genetics, and global systems.

Mr. Mac, as he was widely and affectionately known to his colleagues around the world, was born 100 years ago, four years before the Wright brothers achieved powered flight. His father, a graduate of the University of Alabama, had migrated to Arkansas in 1881 with $3,000 of borrowed capital and successfully established a general store in the small town of Altheimer. His father believed that his children should have a good education and should learn the hard facts of business as early in life as possible. As a result, Mr. Mac had jobs throughout his upbringing and throughout his college years. In one instance, as a teenager delivering newspapers in a lightning storm, the pony he was riding was electrocuted. But in the process of falling dead, his pony threw him far enough away that he was unharmed. Mr. Mac's mother was a deeply devout Southern Methodist and a stern disciplinarian. The story is told that Mrs. McDonnell once took her seven-year-old son to an evangelist and proudly announced, "My son, James, is dedicated to the ministry." It was a secret she had not previously shared with James. Despite that effort by his mother, Mr. Mac's propensities for science and engineering showed up early in his life, as he filled the upper floor of his family's house with wireless telegraphy equipment. Another family story is that one time when a minister came to visit, young James held onto a wire attached to a battery and capacitor behind his back so that when the minister reached out to shake his hand, a giant blue spark jumped the gap, startling the minister and mortifying his mother. Lost in the mists of time is what punishment his mother meted out to young James afterward.

One memorable experience during his childhood was when his mother took the family to the St. Louis World's Fair of 1904. They lived in an inn on the fairgrounds for a full month, and Mr. Mac saw the exposition from corner to corner. This must have helped instill in him his later wide-ranging curiosity. Fifty-five years afterward he reminisced that "the only thing I missed at the fair was the prizefight that my two older brothers sneaked off to one night without taking me. I was mad at them at the time, but I did not mind missing the licking which Mother gave them."

Mr. Mac headed off to his freshman year at Princeton in 1917, where he experienced a growing interest in aviation. His father had constantly urged all of his sons to become professional men, or at least to enter a business that had dignity and potential, unlike aviation. Mr. Mac's two older brothers followed their father's advice. One became a successful architect, the other an eminent banker. Mr. Mac, however, would strike off into the "wild blue yonder" of the fledgling and uncertain world of aviation, in which airplanes were made of wood, cloth, and bailing wire, and pilots took their lives into their hands each time they climbed into the open cockpit. At Princeton Mr. Mac majored in physics; there was no aeronautical engineering department. New horizons opened up for him when he was introduced to the "Life History of Stars" in an astronomy course taught by Henry Norris Russell. In a philosophy course he wrote a major paper about an experiment conducted the previous summer by a British team led by Sir Arthur Eddington to measure the deflection of starlight by the Sun during a solar eclipse, which was the first experimental confirmation of Einstein's general theory of relativity. Also, at the end of his sophomore year he wrote a 32-page manuscript to himself about the mysteries of human existence. He concluded with a note: "I realize that the whole of this is very loosely reasoned. It is meant as simply a sketch, not a philosophical argument. I have written it as an experiment, without reading any philosophy except that required in the sophomore course, to see how the ideas I had in mind compared with those of the philosophers. This comparison is yet to be made." It is unclear whether he ever completed the comparison, but he did continue to ponder those subjects throughout his life. While living in Chicago and working for Western Electric during the summer after his sophomore year, he spent every spare moment in the library reading the works of William James and Frederic W. H. Myers' book, *Human Personality and Its Survival of Bodily Death*. He became convinced that all mental and physical activity, including the so-called paranormal phenomena, had their basis in the performance of the underlying neuronal systems.

After graduating from Princeton with honors in 1921, he entered the only graduate aeronautical engineering program in the country, at the Massachusetts Institute of Technology, as one of only three civilians admitted because the program was organized primarily for Army and Navy officers. Masters degree in hand, he was accepted as an aviation cadet with a commission as a 2nd lieutenant in the Army Air Service Reserve in September 1923. He was one of only six ROTC graduates from across the country that the Army Air Service had enough money to accept for active duty that year.

The next four months he spent in flight training at Brooks Field in San Antonio, Texas. He had a glorious time imitating Jonathan Livingston Seagull, flying the World War I Curtiss Jennys only a few feet over the mesquite, looping, zooming, and trying out all the capabilities of flying.

He also was one of six volunteers who tested the new device known as a parachute. To perform the test he lay on the wing of an airplane as it took off and climbed to altitude. Then on a signal from the pilot he was supposed to let go. On his first try he could not unclench his fingers. Afterward, in his usual analytical style, he marveled at the fact that he could not command his fingers to let go. In any case, on the second try, after the cadet on the other wing had successfully let loose, Mr. Mac let go and found "such quiet, and mental isolation as never experienced on Earth—ecstasy."

While others in his flight class went on to become the leaders of the U.S. Air Force after World War II, 2nd Lieutenant James S. McDonnell wanted to design and build aircraft. So he separated from the Army Air Service and (quoting from a 1973 speech of his), "after finding out he couldn't make a living at it in his native state of Arkansas, he went up amongst the damn Yankees where most of the tiny aircraft plants were, and made out as best he could."

For most of the next 15 years he worked at a series of aircraft companies, including Hamilton, Ford, and Martin. At one point he decided to
strike out on his own with two other engineers to design and build what
would now be called a short-takeoff and -landing airplane. It was all
metal, low wing, and could land on rough fields in about 20 feet. Before it
was even built, he entered it in the Guggenheim Safe Airplane Competition for a $100,000 prize. Airplanes from around the world competed. To
make a long story short, he flew it cross-country to the competition after
only one test flight, arriving just ahead of the qualification deadline. On
the first day, the Doodle-Bug, as he called it, performed spectacularly; but
on the second day the tail broke and Mr. Mac had to crash land it. Although he was among the five finalists, he could not get it repaired in
time to continue in the competition.

He did not give up, however. For the next two years he barnstormed
around the country, trying to get financial backing to produce the Doodle-
Bug as the family flivver of the sky.

In an application for employment as a transport pilot in 1933, he
wrote: ". . . in order to test the feasibility for general private flying of the
'Doodle-Bug,' a new type monoplane with slats and flaps, I flew it some
27,000 miles in 1929-1930 between the Atlantic and Rocky Mountains and
Toronto and Miami, under all such conditions as might be unwittingly
encountered by amateurs: a snow storm, a sleet storm, rainstorms, . . .,
night flying, landing in open country without landing lights or landing
flares on a dark 100% overcast night, several flights with a cloud ceiling of
200 feet cross country, and one flight with a 50 foot cloud ceiling down a
double track railroad, terminated by landing in a visibility of 200 yards;
. . . and landings and take-offs on private lawns and on practice fairways
of golf courses."

Unfortunately, it was the early 1930s, and the country was in a deep
depression. By the end of 1931 he was out of money and went back
to work as an engineer while waiting for the opportunity to start his
company.

Throughout this period he continued to search for greater meaning.
In 1935 he wrote: "What I desire is to find some activity to which I can
devote all of myself and which will lift me out of my small self and enable
me to serve the creative evolution of life on earth as a whole."

But he recognized that to achieve that end he had to have financial
independence. So he consciously rededicated himself to aviation as his
full-time occupation, more convinced than ever that he must start his own
company. Furthermore, he wanted to build commercial transport aircraft
because ". . . swift transport is conducive to world travel and trade and
therefore conducive to the gradual welding of the peoples of the earth
into a more friendly and more harmonious and purposeful community."

He further wrote in 1935: "I feel that a world of flying people will be a world of better people." We may not be better people today, but he was certainly right that the airplane has been instrumental in the globalization of trade and in making those who do travel more broad-minded in their thinking and actions.

His chance to start his own company finally came in the late 1930s, but it was based on a coming war, not on economic growth and prosperity. He left the Glenn L. Martin Company in late 1938, and after an intense eight months of fundraising and preparation, at age 40 he founded McDonnell Aircraft Corporation in St. Louis, Missouri, on July 6, 1939, to build military aircraft. In his usual methodical manner he had picked St. Louis, where he had never previously lived. Because the coming war could make both coasts vulnerable, he wanted a location in the middle of the country. Also, he wanted a good aviation labor market, which St. Louis had. And most important, he was able to attract a considerable number of investors in St. Louis because it had an aviation history, including the financing of Lindbergh's famous flight across the Atlantic in 1927.

During the first year sales were zero, losses were $3,892, shareholders' equity was $218,498, and the company employed about 50 people. Forty years later when he died, annual sales were $6 billion, net earnings were $145 million, dividends paid were $34 million, firm backlog was $8.8 billion, shareholders' equity was $1.5 billion, and the company employed 82,550 people. Furthermore, McDonnell Douglas was the second largest maker of commercial transport aircraft, second in defense contracts, 4th largest U.S. exporter, and 54th largest U.S. industrial corporation. With he and his family owning about 20 percent of the stock, he had certainly achieved his goal of financial independence.

In fact, as soon as the company became successful, he began to expand his personal horizons and to become deeply involved both personally and financially in his lifelong interests outside of building aircraft.

Astronautics and the cosmos are areas that combined both business and pleasure for him. In the mid-1940s he became close friends with nuclear physicist and Nobel Laureate Arthur Holly Compton, who had just become the chancellor of Washington University in St. Louis. He sought out other scientists such as James A. Van Allen, after whom the radiation belts surrounding Earth were named.

He could hardly contain his excitement when McDonnell Aircraft won the contract to design and build the first U.S. manned spacecraft, Mercury, in 1959. He loved interacting with the astronauts and attending the launches.

Although he was not comfortable making public speeches, whenever he did, he meticulously prepared them word by carefully chosen word. Usually they were about space. In 1957 in an address to the graduating class of the University of Missouri engineering school, he declared: "So Fellow Pilgrims, welcome to the wondrous Age of Astronautics. May serendipity be yours in the years to come as Man steps on the Earth as a footstool, and reaches to the Moon, the planets, and the stars."

Mr. Mac put his money where his interests were. In 1963 the city of St. Louis ran out of funding to complete the construction and outfitting of a new planetarium. Mr. Mac donated the necessary funds. In 1963 he funded a Professorship in Space Sciences at Washington University, and in 1975 made a major gift to establish a center for space sciences.

In fact, as early as 1950, he incorporated a private charitable foundation, the McDonnell Foundation (now the James S. McDonnell Foundation), and contributed $500,000 of McDonnell Aircraft stock to it. Adding later contributions and the tremendous increase of McDonnell Aircraft stock value, the foundation's assets have grown to over $300 million today despite $225 million of grants it has awarded during its 49-year history. The Centennial Fellowship Program is only the latest undertaking of the foundation to carry on Mr. Mac's legacy of research and philanthropy.

From 1963 to 1966 Mr. Mac served as chairman of Washington University's Board of Trustees. He used the opportunity to meet with many faculty members and probe with his usual thoroughness into their research. In this way he continued to educate himself in many fields of endeavor.

In the area of genetics he was greatly intrigued and excited when the molecular structure of DNA was discovered by Watson and Crick in the 1950s. He meticulously probed the scientists at Washington University to learn all he could. In 1966 he provided funds for construction of a new medical sciences building, and, looking to the future, he instructed that two extra floors (not needed at the time) be built. He then endowed a new Department of Genetics (to be housed in the extra space), and at the dedication of the building he lectured the audience on the importance of the study of genes for the future well-being of humanity.

In the area of global and complex systems he had two complementary, but seemingly contradictory, interests. He was a great supporter of both NATO, a military alliance, and the United Nations, a global institution to preserve peace. In a 1967 speech he reminded his audience: "As I have said many times, the waging of peace must be achieved from a foundation of great strength. Our mission at McDonnell Aircraft is to contribute to the building and maintaining of that strength." NATO represented the foundation of strength, and the United Nations represented "Man's most noble effort to achieve international peace." As a result, he was actively involved in both organizations and headed the U.S. organization to support the United Nations.

Again, he put his money where his interests were. McDonnell Douglas was the only organization in the world where all employees had paid holidays each year on both NATO Day (April 4) and U.N. Day (October 24).

In 1965 he authored an article titled "Only The United Nations Can" in which he wrote: "The destiny of our Planet, the development of its human and material resources, the spread of scientific knowledge, the population-explosion and the evolving world community—these are the subjects which should make up the continuing dialog between heads of Governments. . . ." He closed the article with the words: "If the United Nations is to fulfill the hopes of all men, the parochial concerns of the moment must be subdued in favor of the long-term universal concerns that ultimately will determine Humanity's future."

Surely the interactions between humans is the most complex system of all. Science and human responsibility was for Mr. Mac another very important theme. In 1957 he funded and participated in a major international meeting on the subject at Washington University. The objective was to examine "How the rapid technological changes in modern life can best be channeled to yield the highest human gains." He returned to the science and human responsibility theme many times in his life, including a 1973 speech of the same title in which he stated ". . . it is only during the past few decades that Man has achieved the science and technology whereby he can either ruin himself and all life on Earth, or he can consciously and responsibly try to help lead all life on Earth further along the path of creative evolution." One of Mr. Mac's favorite artistic depictions of this philosophy was a statue by sculptor Albin Polasek, titled "Man Carving Out His Own Destiny." Mr. Mac kept a replica of the statue in his study at home, and he even had it printed on the McDonnell Douglas stock certificates. It symbolizes Mr. Mac's belief that humanity has the ability to carve out its own destiny—for better or for worse.

In the area of human cognition, Mr. Mac had a lifelong fascination with the workings of the brain and the mind. When he learned of the new positron emission tomography, developed by a team at Washington University, he immediately convened a meeting of Washington University's top researchers to discuss how neuronal activity underlies mental activity and behavior. He challenged the group to prepare a truly innovative research proposal for consideration by the McDonnell Foundation. By the time the proposal was presented in May 1980, Mr. Mac had already suffered the first of a series of strokes from which he died in August of that year. However, he rallied his faculties to further improve the proposal and made a $5 million grant to endow the Center for the Study of Higher Brain Function, which was one of his last acts before passing on to the next phase of existence (as he characteristically referred to death).

I know that Mr. Mac would be tremendously excited by the research projects that are presented in this volume. The recipients of the McDonnell Centennial Fellowships are very fortunate to be involved in research that to paraphrase Mr. Mac's words, "will lift you out of your small self and enable you to serve the creative evolution of life on earth as a whole." Also, with the $1 million grant from the foundation, each fellow will have the financial independence that Mr. Mac believed is so important for maximum creativity.

In closing, I believe that what expressed Mr. Mac's philosophy and approach to life most succinctly and elegantly is a prayer he created and painstakingly improved throughout his lifetime. "Universal Creative Spirit—We thank you for the gift of conscious life on Earth with the opportunity to explore, create, develop, and grow in spirit and the opportunity to nurture all living things and take charge of the creative evolution of same. Hallelujah!"

CARVING OUR DESTINY

1

The Minds of Humans and Apes Are Different Outcomes of an Evolutionary Experiment

Daniel J. Povinelli

Is there anything truly distinctive about the human mind? Charles Darwin did not think so. Indeed, in trying to counter a perceived weakness of his theory, Darwin (1871, p. 445) found it necessary to claim that there was "no fundamental difference" between the mental faculties of humans and other animals. After all, if he could establish that our species did not possess any unique mental features, then there would be less need to invoke a Divine account of our origin. And so, in an extended survey of the animal kingdom, Darwin marshaled an impressive array of anecdotes to defend a claim that he considered self-evident—animals behave like us, therefore they think like us. Rather than viewing it as one kind of mind among many, Darwin presented the human mind as a brighter, more talkative version of other minds; minds that could themselves be arranged in linear order according to how closely they approximated the complexity of the human mind. Darwin's argument fell on the sympathetic ears of George John Romanes, who, shortly after Darwin's death, outlined a method for a new science whose purpose would be to compare the mental structures of humans and animals: "Starting from what I know of the operations of my own individual mind, and the activities which in my own organism they prompt, I proceed by analogy to infer from the observable activities of other organisms what are the mental operations that underlie them" (Romanes, 1882, pp. 1–2). Comparative psychology was thus born of an argument by analogy. Indeed, Romanes' method was

Cognitive Evolution Group, University of Louisiana at Lafayette really just a restatement of what David Hume (1739–1740) had proclaimed a century earlier: Where humans and animals share similar behavior, so too must they share similar minds.

In this chapter, I explore the possibility that Hume, Darwin, and Romanes were wrong. The results of our comparative research program with chimpanzees and human children suggest that the human species may have evolved fundamentally new psychological abilities—abilities that arose not as graded improvements of general cognitive faculties, but as specializations that were woven into existing ancestral neural systems. These innovations in human cognition may have done more than just provide our species with additional psychological capacities. They may have forever altered the way in which we are able to think about and interpret our social behaviors—behaviors that for the most part evolved long before we did. In short, there may be a far more subtle and complex relationship between what we think and how we behave than what was suspected by Hume, Darwin, and Romanes, or the generation of psychologists who followed in their intellectual footsteps.

EVOLUTIONARY HISTORY OF A COGNITIVE SPECIALIZATION

In cultures around the world, humans exhibit a stubborn penchant for explaining behavior in terms of internal mental states (see Avis and Harris, 1991; Lillard, 1998a, 1998b; Povinelli and Godfrey, 1993; Vinden and Astington, 2000). Even in our own culture, where behaviorists admonished us against doing so for the better part of a century, we refuse to stop reflecting on our thoughts, wants, desires, and beliefs, or worrying about similar states in others. But do other species similarly suppose that behavior is a product of an internal set of mental states? Perhaps no other issue strikes so clearly at the heart of what it may mean to be human. Premack and Woodruff (1978, p. 515) framed the question succinctly by asking whether chimpanzees have a "theory of mind." "A system of inferences of this kind," they observed, "may properly be regarded as a theory because such [mental] states are not directly observable, and the system can be used to make predictions about the behavior of others." Premack and Woodruff suggested an affirmative answer to the question of whether chimpanzees possess such a theory of mind, but it is now clear that this is an issue that is far too complex to be resolved by the results of any single experiment. Over the past nine years, we have developed a diverse set of experimental procedures for investigating various facets of theory of mind, and our results point to a very different conclusion compared with the one reached by Premack and Woodruff. Our work points to the possibility that humans have evolved a cognitive specialization for reasoning about mental states.

Some will immediately object to the idea that there are human cognitive specializations, first, by invoking parsimony and, second, by alleging that any such suggestions are anthropocentric. After all, have we not now seen the remarkable behavioral similarities between humans and chimpanzees, such as their abilities to make and use tools, to deceive and manipulate each other, and to recognize themselves in mirrors? And is this not enough to convince us that chimpanzee behavior is governed by an understanding of the social and physical world fundamentally similar to our own?

But before rushing to the conclusion that similarity in behavior implies comparable similarity in mind, we might pause and consider the development of our own minds. Indeed, the most immediate consequence of Premack and Woodruff's (1978) initial report was to excite a flurry of research into the development of young children's understanding of mental states. And, despite the relatively young state of the field, it seems increasingly clear that there are several major transitions in the human understanding of the mental world. No longer can we think of the three-year-old's mind as a duller, incomplete version of the five-year-old's. Rather, psychological development seems to entail successive changes in children's representational capacities so that children of different ages possess different kinds of understandings (e.g., Karmiloff-Smith, 1992). At the same time, observations of their spontaneous interactions with their family and peers reveal a general similarity in their emotions, perceptions, and basic behavioral patterns. Thus, Hume's argument by analogy begins to break down even when we restrict our study to human beings. Worse yet, just as behavioral similarities between children of different ages can obscure psychological differences, so too may behavioral similarities between humans and apes obscure psychological differences.

TOWARD THE SYSTEMATIC STUDY OF THE EVOLUTION OF THEORY OF MIND

In my doctoral dissertation, I reported the results of several experiments I conducted to compare the social intelligence of monkeys, apes, and humans (Povinelli, 1991). But I finished this work with a nagging sense of dissatisfaction. There were few experimental studies of theory of mind in other species, and it seemed that very little progress had been made. In reflecting on the inconclusive nature of this early research, it became apparent that a new approach to comparing human and ape cognition was needed. Several kinds of projects were possible, each with its own strengths and weaknesses. For example, one could take a single chimpanzee, raise it in the company and culture of humans, and test the animal as it developed. Indeed, this has been the traditional approach to studying ape cognition. But would it be possible to maintain the objectivity of such a project? And even if it were, would anyone else be convinced of its objectivity? Finally, would such a project meet one of the central requirements of any scientific undertaking—could it be meaningfully replicated? Another approach was to test large numbers of captive chimpanzees living in social groups. But there were serious problems here as well. For example, without sufficient exposure to humans, how could we be confident that the animals were comfortable enough with our testing procedures that they would perform at the upper limits of their abilities?

In the end, I settled on the following course. I selected 7 two-year-old chimpanzees who had been reared together in a nursery with human caretakers (see Figure 1). These seven animals were transferred to an indoor–outdoor compound that was connected to a specially designed testing facility (see Figure 2). This facility allowed us to test each ape in turn for 10–20 minutes at a time. Thus, while the others played, one of them could be transferred into an outside waiting area, which was connected by a shuttle door to an indoor testing room. Typically, this animal waited outside as a trial was set up indoors. When the shuttle door opened, the animal was free to enter the laboratory and respond to the task. A Plexiglas panel separated the apes from the humans. We used this panel for two reasons, even when the apes were young enough that it was unnecessary. First, it offered unambiguous response measures for many of our tests (i.e., which hole they reached through); but more important, we wanted to establish a predictable setting and routine that we could use with these animals throughout their youth and later into their adolescence and adulthood as they reached their full size and strength. That this approach has paid off is clear from the fact that as young adults our animals still eagerly participate in testing two to three times a day. To be sure, our project has had its limitations, but as will become clear, it has had its unique set of strengths as well.

This, then, was our approach: Rear a cohort of chimpanzees together, while simultaneously exposing them to human culture; follow them through their juvenile years into adulthood, and compare their social understanding to that which develops in human children. However, such work would require as fair a set of comparisons with young children as possible, and so we established a center where similar experiments with young children could be conducted. And although I do not discuss it in this chapter, this center has also allowed us to probe other, underexplored aspects of children's developing understanding of themselves in time—developments that may ultimately prove crucial in defining both the similarities and the differences between humans and apes.

For the past nine years, our cohort of seven apes has participated in dozens of experiments investigating their understanding of mental states

FIGURE 1 Seven chimpanzees (Group Megan) who have formed the basis for our long-term project exploring the nature of chimpanzees' reasoning about the social and physical world. These chimpanzees began this project when they were 30–42 months old. They were between the ages of four and five years old when these photographs were taken.

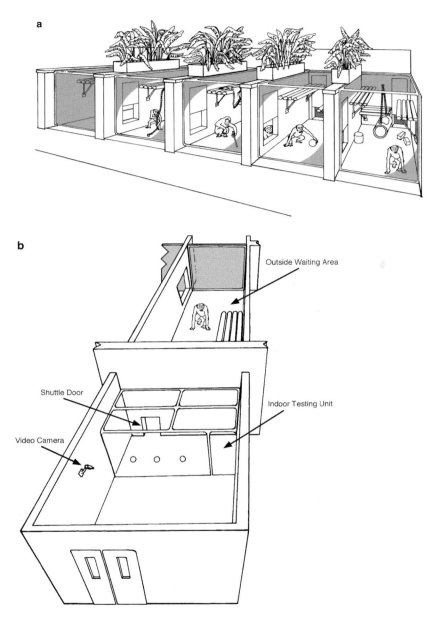

FIGURE 2 (a) The living area for Megan and her peers showing their indoor–outdoor living areas; and (b) close-up diagram of outdoor waiting area and indoor test unit showing the shuttle door connection, Plexiglas partition, and experimental working space.

such as attention, intention, desire, knowledge, and belief, as well as their understanding of self (see Povinelli and Eddy, 1996a, 1996b, 1996c, 1997; Povinelli and O'Neill, 2000; Povinelli et al., 1993, 1994, 1997, 1998, 1999; Reaux et al., 1999; Theall and Povinelli, 1999). Where informative, we have also tested large numbers of other chimpanzee subjects (from infants to adults). In concert with our work with young children, the results of these studies have allowed us, in the very least, to gain a clearer perspective on the question that motivated me to pursue this research in the first place: Are there unique aspects of human cognition, and if so, how do they emerge from among the complex set of developmental pathways that we share in common with chimpanzees and other primates?

UNDERSTANDING VISUAL PERCEPTION: A CASE STUDY

The Mind of the Eyes

Although we have investigated and compared numerous aspects of chimpanzees' and children's understanding of self and other, we have focused much of our effort on determining whether apes (like us) interpret the eyes as windows into the mind. In short, one of the main agendas of our work has addressed what chimpanzees know about "seeing." This is critical for questions concerning the evolution of theory of mind because of the importance we place on the eyes in our understanding of attention and knowledge. In one sense, the eyes occupy center stage in human folk psychology. Humans understand the perceptual act of seeing as far more than just a physical relation between eyes and objects and events in the world, and this understanding arises fairly early in development. For example, a four-year-old girl observes someone lift the lid of a box and look inside and automatically infers that the person *knows* its contents. In contrast, she does not make this assumption when someone merely touches the box. Thus, even though she cannot see the information entering a person's eyes and then forming the mental state of knowing the box's contents, she (like an adult) assumes that something very much like that has happened. This may seem trivial, until we try the same test with three-year-olds and discover that they appear oblivious to the psychological difference between the person who looked into the box and the person who only touched it (Gopnik and Graf, 1988; O'Neill and Gopnik, 1991; O'Neill et al., 1992; Perner and Ogden, 1988; Povinelli and deBlois, 1992b; Ruffman and Olson, 1989; Wimmer et al., 1988).

Does this mean that three-year-olds have no understanding of the mentalistic aspects of visual perception? Not necessarily. Even though they may not appreciate that seeing (or other forms of perception) causes internal mental states, they do seem to understand that seeing is "about" (or refers to) events and objects in the external world. In short, they seem to realize that someone who is looking at an object or event is subjectively connected to it by something that we label "attention." Many years ago, Flavell and his colleagues conducted an impressive series of studies that revealed at least two clear levels in the development of preschoolers' understanding of seeing (see Flavell et al., 1978, 1980, 1981; Lempers et al., 1977; Masangkay et al., 1974). The first level is exhibited by children aged two to three who possess a firm understanding of whether or not someone can see something. For example, sit across a table from a child and show the child a picture of a turtle so that it is right side up from the child's perspective, but upside down from yours. Three-year-olds know that you can see the turtle, and that if you put your hands over your eyes you can no longer see it. But they do not seem to understand seeing at a second, more sophisticated level: They fail to understand that, from your perspective, the turtle appears different (i.e., upside down).

Even infants display some knowledge about visual gaze. In a landmark report over two decades ago, Scaife and Bruner (1975) demonstrated that very young infants will turn and look in the direction they see someone else looking. Since then, a number of carefully controlled studies have explored the emergence of gaze following (or, as Scaife and Bruner called it, "joint visual attention"). Although there is disagreement about the exact timing of its development, some capacities related to gaze following may emerge as early as six months (Butterworth and Cochran, 1980; Butterworth and Jarrett, 1991; Corkum and Moore, 1994; Moore et al., 1997). By 18 months of age, however, the ability is well consolidated, as toddlers will (a) follow an adult's gaze into space outside their own visual field, (b) precisely locate the target of that gaze, and (c) reliably follow the gaze in response to eye movements alone (without accompanying movements of the head).

Although gaze following is of obvious practical utility, its psychological standing remains less clear. Some, such as Baron-Cohen (1994) of Cambridge University, interpret gaze following as prima facie evidence that infants are explicitly aware of a psychological connection between self and other. In other words, infants turn to follow their mother's gaze because they know that she is looking *at* something, that she *sees* something, that something has engaged her *attention*. This account grants infants their first (albeit limited) glimpse into the visual psychology of other people. Other researchers are more cautious, maintaining that gaze following (especially in very young infants) may have little to do with an appreciation of internal psychological states (Butterworth and Jarrett, 1991; Moore, 1994; Povinelli and Eddy, 1994, 1996a, 1996b; Tomasello, 1995). Processes such as hard-wired reflex systems, learned behavioral contingencies, and attentional cuing have all been offered as possible explanations of the emergence of this behavior. There is, finally, a middle ground that interprets the behavior as a causal precursor to a later-emerging, more explicit representation of attention in toddlers. Early gaze following is seen as a fairly automatic response, which later provides a context for developing an understanding that attention is something distinct from action itself.

Gaze following is not the end of the story. Our information processing about the eyes of others is connected to emotional as well as cognitive systems. Making eye contact, for example, is a highly significant emotional experience for infants long before they appreciate the attentional aspect of seeing (Hains and Muir, 1996; Symons et al., 1998; Wolff, 1963). And it is not just human infants who display such sensitivity. Stare into the eyes of many species of Old World monkeys and note their immediate, hostile reaction (Perrett et al., 1990; Redican, 1975). In contrast, mutual gaze plays a more flexible role in humans and other great apes. Here, mutual gaze is an important factor in mediating both agonistic and affiliative social interactions (de Waal, 1989; Gómez, 1990; Goodall, 1986; Köhler, 1927; Schaller, 1963). In chimpanzees, for example, establishing mutual gaze seems to be especially important during "reconciliatory" social interactions that immediately follow conflicts (de Waal, 1989). In general, sensitivity to the presence of eyes appears to be widespread among animals, presumably because the presence of a pair of eyes is linked with the presence of a predator. In a well-known example, Blest (1957) studied the function of so-called "eye spots" on the wings of butterflies. He experimentally demonstrated that butterflies with such spots are less likely to be eaten by predatory birds (see also Burger et al., 1991; Burghardt and Greene, 1988; Gallup et al., 1971; Ristau, 1991).

But does the evolved sensitivity to the presence, direction, and movement of the eyes indicate an understanding of their connection to internal mental states, even simple ones such as attention? Does the bird who averts from striking a butterfly after being flashed a set of eye spots entertain the notion of having been "seen"? In this case, our intuition may reply, "no," but perhaps only because birds are involved. And what about gaze following in human infants, or possibly other species? Gaze following inhabits a contentious middle ground where intuitions clash. Clearly, then, we need to move beyond intuition and instead ask whether we can experimentally disentangle alternative psychological accounts of gaze following. In this way, we can begin to make real progress toward understanding whether other species (especially those most closely related to us) reason about visual perception in the mentalistic manner that we do.

Gaze Following: First Steps Toward Understanding Attention?

To begin, we might ask if chimpanzees possess a gaze-following system that is homologous to that of humans, and if so, the extent of similarity between the two. Following suggestive accounts from field researchers (e.g., Byrne and Whiten, 1985), several years ago we conducted the first experimental studies of gaze following in nonhuman primates (see Povinelli and Eddy, 1996a, 1996b, 1997; Povinelli et al., 1999). Initially, we simply sought to determine whether chimpanzees follow gaze at all. Our procedure involved our subjects entering the laboratory and using their natural begging gesture to request an apple or banana from a familiar experimenter who sat facing them. On most trials, the experimenter immediately handed the subjects the food. On probe trials, however, the subjects were randomly administered three conditions that had been carefully choreographed ahead of time. On *control* trials, the experimenter looked at them for five seconds and then handed them a reward as usual. These trials allowed us to determine the subjects' baseline levels of glancing to specific locations in the room. On *eyes-only* trials, the experimenter turned his or her head and looked at a target above and behind the chimpanzee for five seconds (see Figure 3a). On *eyes-only* trials, the experimenter diverted only his or her eyes to the same target, keeping the rest of the head motionless. The results depicted in Figure 4 reveal several things. First, the chimpanzees virtually never looked above and behind themselves on the control trials. In clear contrast, not only did they follow our gaze on the trials involving whole head motion, they even did so in response to eye movements alone. We have now replicated and extended this effect on a number of occasions and have demonstrated that chimpanzees follow gaze with at least the sophistication of 18-month-old human children (see Table 1).

These findings were important because they experimentally established that chimpanzees, like humans, are extremely interested in where others are looking. More recently, researchers in Perrett's laboratory at the University of St. Andrews, and Tomasello's laboratory at Emory University, confirmed our speculation that gaze following might be widespread among primates (Emery et al., 1997; Tomasello et al., 1998). Indeed, it seems likely that gaze following evolved through the combined effects of dominance hierarchies (needing to keep track of who is doing what to whom) and predation (exploiting the reactions of others to discover the location of potential predators). Having established this similarity, we next turned to the more central question of how this similarity might help us to understand whether chimpanzees interpret visual perception as a projection of attention.

Knowing That You Cannot See Through Walls

A chimpanzee following your gaze tugs almost irresistibly at a natural inclination to assume that he or she is trying to figure out what you have just seen. But what excludes the possibility that they are simply look-

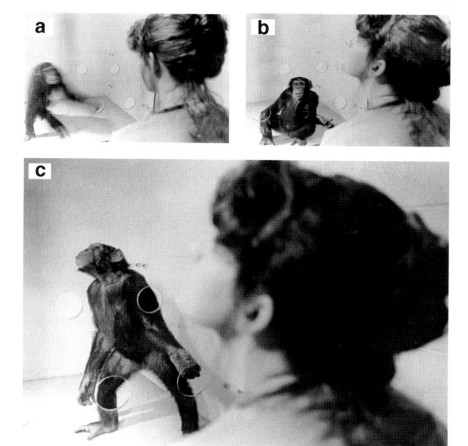

FIGURE 3 Gaze following in a five-year-old chimpanzee in response to eyes and head condition.

ing where you are looking, without entertaining ideas about your internal attentional state? To intelligently choose between these different accounts of gaze following, we need to flesh them out a bit more clearly and determine if they generate different predictions about how apes (and human infants) might respond in more revealing circumstances.

In a first account, we envisioned that chimpanzees and other nonhuman primate species (and even human infants) understand "gaze" not as a projection of attention, but as a directional cue (i.e., a vector away from the eyes and face). Thus, perhaps the ancestors of the modern primates merely evolved an ability to use the head–eye orientation of others to

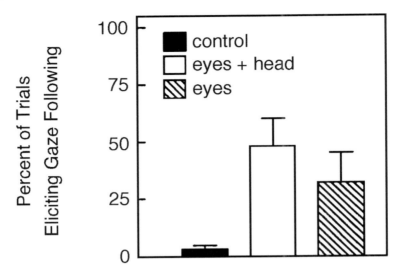

FIGURE 4 Results of the initial study of gaze following in juvenile chimpanzees.

move their own visual systems along a particular trajectory. And, once their visual system encountered something novel, the operation of the primitive orienting reflex would ensure that two chimpanzees, for example, would end up attending to the same object or event, without either

TABLE 1 Behavioral Evidence that Humans and Chimpanzees Possess a Homologous Psychological System Controlling Gaze Following

Behavior	18-Month-Old Human Infants	Juvenile and Adult Chimpanzees
Respond to whole head movement?	yes	yes
Respond to eye movement alone?	yes	yes
Left and right specificity?	yes	yes
Follow gaze outside immediate visual field?	yes	yes
Scan past distractor targets?	yes	yes
Account for opaque barriers?	?	yes

SOURCES: Butterworth and Cochran (1980), Butterworth and Jarrett (1991), Call et al. (1998), Povinelli and Eddy (1996a, 1996b, 1997), Povinelli et al. (1999), Tomasello et al. (1998).

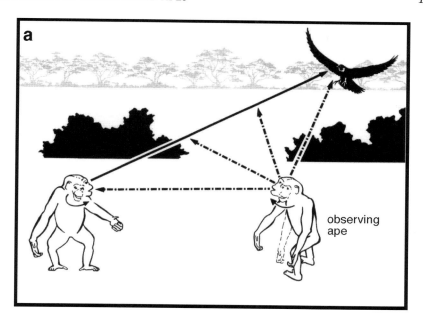

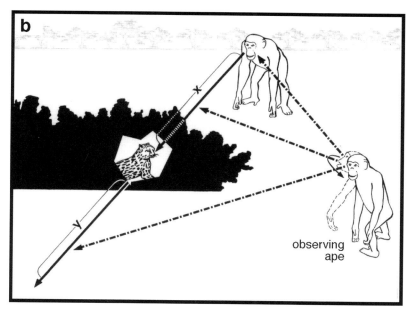

FIGURE 5   Natural contexts for reasoning about the visual gaze of others: (a) exploiting the gaze of others to discover important events and (b) natural scenario that might distinguish between high- versus low-level models of gaze following.

ape representing the other's internal attentional state (Figure 5a). In contrast to this psychologically sparse interpretation, we considered a second account: Apes might follow gaze because they appreciate its connection to internal attentional states. Although I fear being misunderstood, it is necessary to have a shorthand label for these alternative kinds of accounts, and so I hereafter refer to them as the "low-level" and "high-level" accounts, respectively. The so-called high-level model stipulates that chimpanzees form concepts about internal mental states (such as attention) and use these concepts to help interpret the behavior of others. In contrast, the low-level model supposes that chimpanzees cogitate about behavioral propensities, not mental states.

In considering how to distinguish between these explanations of gaze following, it occurred to us that if nonhuman primates reason about the attentional aspect of gaze, this might reveal itself most clearly when they witness another animal's gaze being obstructed by an opaque barrier such as a tree, or in the case of captivity, a wall. If the high-level account of their gaze-following abilities were correct, and if an observing animal were to witness another animal in the situation depicted in Figure 5b, the observing animal should be capable of understanding that the other ape cannot see through the obstruction. If so, the observing ape should look around the barrier to determine what the other ape had seen. In contrast, the low-level account predicted that the observing chimpanzee would project a vector away from the other ape's face and scan along this path until something novel triggered an orienting reflex, and, if nothing novel were present, stop scanning altogether.

We tested these possibilities in our laboratory (see Figure 6a–b; Povinelli and Eddy, 1996b). To begin, we covered half of the Plexiglas panel with an opaque partition. Thus, the chimpanzees could still enter the test lab, approach an experimenter, and request some fruit by begging through a hole in the Plexiglas. However, the partition blocked the apes' view into a small area of the room behind the partition—an area into which only the human could see. On most trials, the subjects simply entered the lab, gestured to the human, and were handed the food. However, several experimental conditions allowed us to test the accounts described above. In one condition, as soon as the chimpanzee reached through the Plexiglas, the experimenter looked at the subject while executing a choreographed series of irrelevant movements for precisely five seconds. This condition (along with several others) allowed us to measure the subject's ambient levels of glancing to various locations in the room (Figure 6a). In contrast, on the crucial test trials the experimenter leaned and glanced at a predetermined target on the front of the partition (see Figure 6b). Would the apes attempt to look around the partition to the experimenter's side, or would they follow a vector away from the

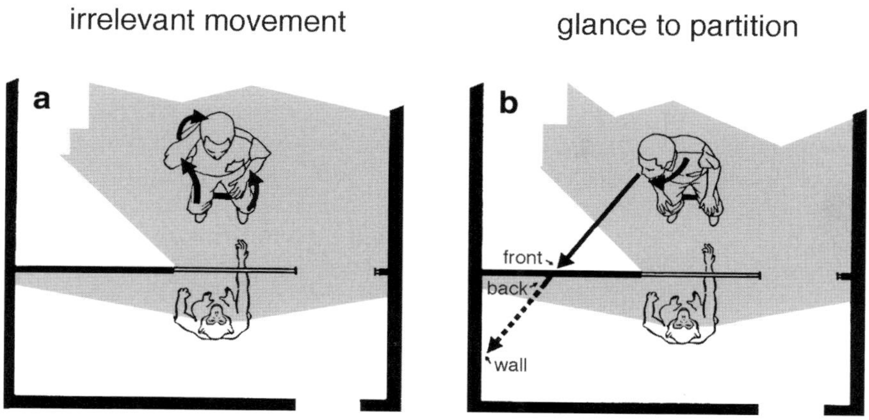

irrelevant movement

glance to partition

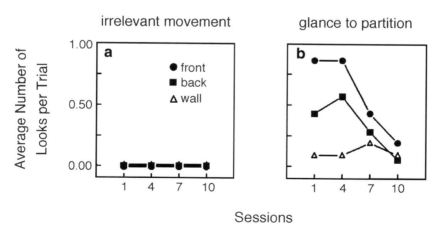

FIGURE 6 Laboratory recreation of a natural context (see Figure 5b) for investigating whether chimpanzees understand that gaze cannot pass through opaque barriers.

experimenter's face until they wound up looking behind themselves at the back wall of the test lab? In short, would they (at some level) process the fact that the experimenter's vision was blocked by the partition? The accounts described above generated very different predictions about the answer our apes would provide.

As can be seen in Figure 7, the chimpanzees' reactions were unambiguous—they behaved in the manner predicted by the high-level expla-

FIGURE 7 Effect of experimenter's movements on where chimpanzees looked on the opaque barrier test (see Figure 6).

nation. Instead of blindly turning their heads toward the back of the test unit, the apes looked around to the front of the partition (Figure 7b), exactly as if they understood that the experimenter could not see through it. Thus, despite the fact that the experimenter was never really glancing at anything, and hence the apes were never reinforced for looking around the partition, they nonetheless did so on both the first and the second trials, until gradually stopping altogether. In contrast, they almost never looked around the partition when the experimenter merely engaged in irrelevant movements (Figure 7a).

These results troubled us, but not because we were committed to low-level models of chimpanzee cognition. Rather, prior to conducting these gaze-following studies, we had conducted over a dozen experiments exploring whether our apes had a concept of "seeing"—experiments that suggested an answer very different from the high-level account of gaze following (see Povinelli, 1996; Povinelli and Eddy, 1996a; Reaux et al., 1999). In what follows, I summarize these experiments and explore whether they are truly incongruent with the results of our gaze-following studies and, indeed, whether we really need the high-level model to explain our apes' abilities to "understand" that gaze cannot pass through opaque barriers.

Knowing Who Can See You

Our first approach to asking our apes about "seeing" was to determine if they understood the psychological distinction between someone who could see them and someone who could not. To answer this, we again capitalized on one of their most common communicative signals— the begging gesture. Chimpanzees use this gesture in a number of different communicative contexts. For example, a chimpanzee may see you walk past his or her compound with some food and immediately reach toward you with his or her palm up (to "request" the item, as it were). This seemed like an ideal, natural context in which to ask our animals whether the deployment of their gestures was mediated by a notion of "seeing."

We began by training the apes to enter the laboratory and gesture through a hole directly in front of a single, familiar experimenter who was either standing or sitting to their left or right. On every trial that they gestured through the hole directly in front of the experimenter, this person praised them and handed them a food reward. In short order, the apes were all reliably gesturing through the correct hole toward the experimenter (Figure 8a–c).

Of course, we already knew that chimpanzees were inclined to direct their begging gesture toward us. What we really wanted to know was

FIGURE 8 Chimpanzee (a) enters indoor test unit, (b) uses species-typical begging gesture to "request" food from experimenter, who (c) hands subject a piece of fruit.

how the chimpanzees would react when they entered the test lab and encountered not a single experimenter, but two experimenters, one who could see them (and therefore respond to their gestures) and one who could not. With this in mind, we created several clear cases (at least from our human point of view) of *seeing* versus *not seeing*. Although we carefully designed and rehearsed these scenarios, we did not simply pluck them out of thin air. Rather, we studied our animals' spontaneous play and modeled our scenarios after several of their natural behaviors. At the time, one of their favorite pastimes was to use objects in their enclosure to obstruct their vision. For example, they would place large plastic buckets over their heads and then carefully move around their compound until they bumped into something. Occasionally, they would stop, lift the bucket to peek, and then continue along on their blind strolls. Although buckets were the favored means of obstructing their vision, they also used plastic bowls, burlap sacks, pieces of cardboard, and even their hands to produce the same effect. From a common-sense point of view, it seemed obvious that they knew exactly what they were doing—preventing themselves from seeing. In the end, we tested the animals using the conditions depicted in Figures 9a–d in sessions of 10 trials. Eight of these trials were of the easy variety (with a single experimenter). The other two were critical probe trials of the seeing and not seeing conditions.

So how did the animals react when they encountered two familiar experimenters, for example, one blindfolded, the other not? They entered the lab, but then (measurably) paused. And yet, having apparently noted the novelty of the circumstance, they were then just as likely to gesture to the person who could *not* see them as to the person who could. The same effect occurred on the buckets and hands over the eyes conditions: The chimpanzees displayed no preference for gesturing toward the experimenter who could see them. In contrast, on the easy surrounding trials, the apes gestured through the hole directly in front on the single experimenter on 98 percent of the trials. Thus, despite their general interest and motivation, when it came to the seeing and not seeing conditions, the animals appeared oblivious to the psychological distinction between the two experimenters.

There was, however, one exception. Unlike the blindfolds, buckets, and hands over the eyes trials, on back versus front trials (in which one person faced toward the ape and the other faced away; see Figure 9a) the animals seemed to have the right idea: "Gesture to the person who can see." But why the discrepancy? Why should the apes perform well from their very first trial forward on a condition in which one of the experimenters was facing them and the other was facing away, but then not on any of the other conditions? In defense of the high-level account, it could be that the back and front condition was simply the easiest situation in

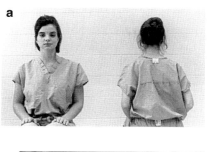

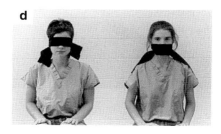

FIGURE 9 Conditions used to explore chimpanzees' understanding of "seeing" versus "not seeing": (a) back and front, (b) buckets, (c) hands over eyes, and (d) blindfolds.

which to recognize the difference between seeing and not seeing. Despite the fact that the animals had measurably paused before making their choices in the other conditions as well, and despite the fact that we had observed them adopt these other postures in their play, the idea that back and front was just a more natural distinction seemed plausible. But there was another, more mundane potential explanation of the results. Compare Figures 8a–c with Figure 9a. Perhaps on the back and front trials the apes were merely doing what they had learned to do—enter the test lab, look for someone who happened to be facing forward, and then gesture in front of him or her. Rather than reasoning about who could see them, perhaps the apes were simply executing a procedural rule that we had inadvertently taught them during testing.

At this point there were several ways of distinguishing between these possibilities. If the high-level account were correct (that is, if the back and front condition was simply the most natural case of seeing and not seeing), then the apes ought to perform well on other, equally natural conditions. Here is a situation that our apes experience daily. One ape approaches a group of others who are facing away from her. As she gets closer, one of the other apes turns and looks at her. Now, although she notices this behavior, does she understand that this other ape is *psycho-*

logically connected to her in a way that the others are not? The new condition that this consideration inspired ("looking over the shoulder," Figure 10) was of interest in its own right, but we had an even stronger motivation for testing the apes on such a condition. Recall that the low-level account could explain our apes' excellent performance on the back and front condition by positing that they were simply being drawn to the forward-facing posture of a person. But in this new, looking over the shoulder condition, there was no forward-facing posture—just the face of one experimenter and the back of the other one's head. Thus, the low-level account generated the seemingly implausible prediction that the apes would perform well on the back and front condition, but randomly on the looking over the shoulder trials.

To our surprise, this is exactly what the apes did. In full support of the low-level model, on the looking over the shoulder trials the apes did not prefer to gesture to the person who could see them. In direct contrast, they continued to perform without difficulty on the back and front trials. This result made a deep impression on us. No longer was it possible to dismiss the results of the other conditions by supposing that the animals thought we were peeking from under the buckets or blindfolds, or between our fingers. No, here we had made "peeking" clear and explicit, and yet the apes still performed according to the predictions of the low-level model. The experimental dissection of the front of the experimenter from the face using a posture they must witness every day (looking over

FIGURE 10 Looking over the shoulder condition.

the shoulder) sobered us to the possibility that perhaps our apes genuinely might not understand that the experimenters had to *see* their gesture in order to respond to it. Furthermore, the results seemed to imply that even for the back and front condition, the apes might have no idea that the experimenter facing away was "incorrect"—rather, someone facing away was simply a posture with a lower valence. They were, after all, perfectly willing to choose the person in this posture on fully half of the looking over the shoulder trials.

We had difficulty accepting the implications of these results. After all, we had witnessed our apes using their begging gesture on hundreds of occasions. Was it really possible that a gesture so instantly recognizable to us could be understood so differently by them? Although their behavior continued to be best explained by the low-level model, chimpanzees are alert, cognizing organisms, extremely attuned to the behaviors that unfold around them. So it would be truly surprising if they failed to learn anything after repeated experience on our tests. To be fair, we had intentionally kept the number of test trials in each experiment to a minimum (typically four) in order to minimize their rate of learning. After all, at this point we were interested in what they understood, not in what they could learn through trial and error.

Nonetheless, we were skeptical of our results and decided to examine their reactions to several other conditions, such as one involving screens (Figure 11). Before testing, we familiarized the subjects with these screens

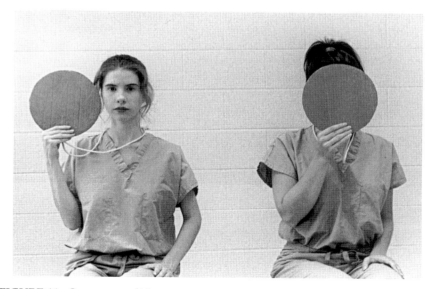

FIGURE 11   Screens condition.

by playing games in which we held the screens in front of our faces, or we let the apes play with the screens themselves. Despite this, when it came to testing, the apes were just as likely to choose the person who could not see them as the person who could. However, after several additional experiments involving the screens condition, their performances began to improve, until finally they were reliably gesturing to the person who could see them. Indeed, follow-up studies revealed the interesting (although not completely unexpected) fact that our apes' correct responding had generalized from the screens condition to the looking over the shoulder condition as well.

At this point, it was possible to walk away from these studies concluding that the apes had simply learned another procedural rule: "Gesture in front of the person whose face is visible." However, nothing seemed to eliminate the possibility that, although they did not do so immediately, they might have finally figured out what we were asking them—"Oh! It's about *seeing*!" We devised several additional procedures for distinguishing between these possibilities. First, we administered the original set of conditions to the apes (blindfolds, buckets, etc.). The high-level account predicted that, because they had finally learned that the task was about seeing, the apes would perform excellently on all of them. The low-level account also predicted excellent performance—except in the blindfolds condition (where blindfolds covered the eyes of one person versus the mouth of the other). Why the blindfolds condition? Because in this condition, an equal amount of the face of each person was visible (see Figure 9d). Although it is perfectly obvious to us that one person in this condition can see and the other cannot, if our apes had merely acquired a set of arbitrary procedural rules about the presence or absence of a face, they would be forced to guess who was correct, choosing the person whose eyes were covered as often as the person whose mouth was covered. And, to our surprise, that is precisely what they did.

Another test confronted the chimpanzees with two experimenters whose eyes and faces were both clearly visible. However, only one of them had his or her head directed toward the ape. The other appeared (to us) distracted, with her head directed above and behind the chimpanzee (see Figure 12). The high-level model predicted that the apes would gesture to the experimenter who was visually attending, whereas the low-level model, because the eyes and faces were present in both cases, predicted that they would gesture at random. The results were striking. The apes entered, looked, and then followed the distracted experimenter's gaze into the rear corner of the ceiling. Nevertheless, the subjects were then just as likely to gesture to the distracted person as toward the person who was looking in their direction. It was as if they processed the infor-

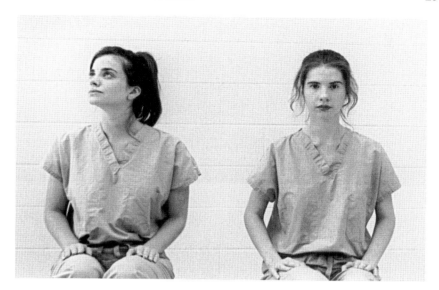

FIGURE 12   Distracted versus attending condition.

mation about the distracted experimenter's direction of gaze, without interpreting its attentional significance.

Finally, we tested the apes on the most subtle version of this task we
could imagine: eyes open versus closed. Although they initially had no
preference for the person whose eyes were open, after a number of trials
their performance improved. However, even here, additional control tests
revealed that when the eyes and face were pitted against one another (see
Figure 13), the face rule was more important! In short, through trial and
error learning (probably aided by an innate sensitivity to the face, eyes,
and overall posture of others) our apes appeared to have learned a hierar-
chically organized set of procedural rules: (1) gesture to the person whose
front is facing forward, if both fronts are present (or absent); (2) gesture to
the person whose face is visible, if both faces are visible (or occluded); (3)
gesture to the person whose eyes are visible. "*Seeing*," then, did not appear to be a concept recruited by the chimpanzees to help them decide to
whom they should gesture.

Validating the Task

In reflecting on these results, we considered the possibility that we
had underestimated the difficulty of our task. It was possible that our
chimpanzees might understand the attentional aspect of seeing, but that

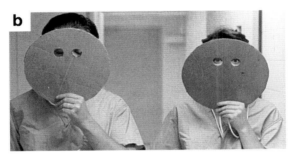

FIGURE 13 Two of the conditions used to distinguish rules used by apes: (a) face versus eyes, (b) eyes versus no eyes, and (c) Mindy gesturing to the correct experimenter.

our task might require a more sophisticated understanding of visual perception than we had thought (for example, the connection between seeing and knowing). Worse yet, our task might be tapping into capacities unrelated to the question of seeing. Fortunately, findings from several laboratories had converged to suggest that an understanding of seeing as attention is beginning to be consolidated in young children by about 30 months of age (Gopnik et al., 1995; Lempers et al., 1977). Thus, we reasoned that if our tests were measuring an understanding of the attentional aspect of seeing, then 30–36-month-old children ought to perform quite well on them. On the other hand, if the tasks required an understanding of the connection between seeing and knowing (which develops at around four years of age in human children), then younger children should perform poorly. We investigated this by training two-year-, three-year-, and four-year-old children, over a three- to five- week period, to gesture to familiar adult experimenters to request brightly colored stickers (see Povinelli and Eddy, 1996b, Experiment 15). We then tested them on several of the conditions we had used with the apes (screens, hands over the eyes, and back versus front). Unlike the apes, the children were correct in most or all of the conditions from their very first trial forward—even the majority of the youngest ones that we tested.

Longitudinal Reflections

All of this may seem puzzling, especially given my above account of how the chimpanzees seemed to enjoy obstructing their own vision. Were there crucial methodological limitations in our seeing and not seeing tests that somehow prevented our apes from displaying an (existing) understanding of seeing? Although possible, consider the following example. An ape feels an irritation on its arm and scratches it. The explanation of this action seems simple: The animal produced a behavior (scratching) that is associated with a reinforcing experience (the cessation of itching). And, as much as our folk psychology resists the idea, the "peekaboo" games we observed our apes playing can be explained in a similar manner: They place a bucket over their heads because it produces an interesting, pleasurable experience. Thus, the *experience* of visual occlusion need not be represented any more explicitly than any other sensation (e.g., the soothing that results from scratching). Such an account could reconcile the seemingly incongruous aspects of our data: Our animals' natural ability to produce visual deprivation in their play behavior alongside their bemusement when asked to explicitly reason about such visual deprivation in others.

But a more troubling question remained. By participating in our tests, our apes had clearly learned relationships about their gestures, on the one hand, and the experimenters' postures, faces, and eyes, on the other. Indeed, they had learned them so well that they could rapidly generalize this understanding to new conditions. This left us in a difficult quandary: How were we to distinguish between the following, very different explanations of these effects? One possibility was that the subjects had not initially possessed a concept of visual attention (or only a weak one), but after extensive experience with our tests they constructed the concept (or at least learned how to apply it in this particular context). But a second possibility refused to quietly go away. Perhaps the subjects neither entered nor emerged from our tests with such a concept, but rather, as a consequence of the differential reinforcement they received, they relearned the set of procedural rules described above—rules that allowed them to behave exactly as if they were reasoning about seeing.

Fortunately, at least one way of distinguishing between these possibilities presented itself 13 months after we completed the studies described above. In the context of a different protocol, we readministered several of the original conditions that our apes had learned very well (e.g., screens, buckets; see Povinelli, 1996). But despite what they had learned a year earlier, the animals responded at chance levels! Indeed, it was only after 48 massed trials of the screens condition that the animals' performance began to creep up to levels significantly above chance.

Although a retention of performance would not have been particularly informative, their failure to do so was. Consider again the history of the children who participated in our tests. Surely they had received numerous semistructured experiences of "seeing" versus "not seeing" before coming into our lab (so many so, in fact, that their first trials were more equivalent to our apes' final trials). However, imagine them returning to our lab a year later. The contrast between the enduring understanding that they had apparently constructed during their second year of life, and the failure of our apes to retain what they had learned a year earlier, suggests in a rather dramatic manner that despite superficial similarities in their performances, the two species might nonetheless have parted company conceptually. Indeed, this distinction is even more sobering when one considers that our apes were not idle during this intervening year. On the contrary, they participated in numerous studies that explored their understanding of attention in other ways. One reading of these results is that far from serving to assist them, at some level these other studies may actually have interfered with their performance on this task.

We were so struck by the apes' weak retention that when they turned eight to nine years of age, we administered a final series of experiments using these same procedures (see Reaux et al., 1999). Again, however, the apes initially responded randomly. We decided to force the issue by training the subjects to perform at above-chance levels. Although not immediately, the apes *did* relearn how to solve many of these problems within about 16 trials. Indeed, some of our chimpanzees learned to solve them all. Did this mean that now, finally, they had constructed an understanding of seeing?

We considered two conditions that several of our apes had learned extremely well: eyes open and closed and looking over the shoulder. We realized that by mixing these conditions together—that is, by taking the *correct* posture from the looking over the shoulder condition and combining it with the *incorrect* posture from the eyes open and closed condition (see Figure 14)—we could pit the two alternatives against each other. After all, if the apes had truly constructed a concept of seeing, then they would be expected to perform excellently in this new, mixed condition. But if, instead, they had simply learned a set of hierarchical rules concerning the fronts, faces, and eyes of the experimenters, then not only would they *not* prefer the (correct) person looking over her shoulder, they would display a significant preference for the (incorrect) person facing forward with her eyes closed. This is because the rule-based model that had so neatly accounted for our apes' previous behavior explicitly stipulated that the forward-facing posture was more important than whether the eyes were open or closed. The outcome was remarkable. Despite their contin-

FIGURE 14 Mixed condition combining *incorrect* option from eyes open versus closed and *correct* option from looking over the shoulder.

ued excellent performance on the eyes open and closed and looking over the shoulder conditions, when we mixed these treatments together as described above, just as the low-level model predicted, the apes gestured to the incorrect experimenter—on every trial!

Gaze Following and "Seeing:" Toward a Reconciliation

Although we were impressed by the utility of the low-level model in predicting our apes' behavior on the seeing and not seeing tests, we pondered the model's apparent underestimation of their ability to appreciate that gaze cannot pass through opaque partitions. One possible reconciliation was to question the generality of the seeing and not seeing tests. For example, perhaps the apes just had trouble simultaneously reasoning about the different visual perspectives of two persons, or had difficulty understanding themselves as objects of visual attention. With these ideas in mind, we explored whether they would show better evidence of understanding the attentional aspect of visual perception in situations more directly involving their gaze-following abilities (see Povinelli et al., 1999). First, we taught our apes, and three-year-old children, to search under two opaque cups for a hidden treat. Next, we occasionally kept them ignorant as to the treat's location, but instead let them witness the experimenter gazing either at the correct cup (at target) or above the correct cup (above target) (see Figure 15a and 15b). We reasoned that if the subjects understood the referential significance of the gaze of the experimenter, they ought to select the correct cup on the at-target trials, but should choose *randomly* between the two cups on the above-target trials. The latter prediction is the key one, because organisms with a theory of attention (for example, human children) should interpret the distracted experimenter as being psychologically (attentionally) disconnected from the cups—conveying no information about the location of the reward.

As expected, the three-year-old children selected the cup at which the experimenter was looking on the at-target trials, but chose randomly between the two cups on the above-target trials. This result provided crucial evidence that our theory of what the task was measuring was correct. In direct contrast, however, the chimpanzees did not discriminate between the at-target and above-target trials. Rather, they entered the test unit, moved to the side of the apparatus in front of the experimenter's face, and then chose the cup closest to them. Did the apes simply not notice the direction of the experimenter's gaze on the above-target trials, thereby confusing these with the at-target trials? Hardly. They followed the experimenter's gaze by looking above and behind themselves on over 71 percent of the above-target trials as compared with only 16 percent of the

FIGURE 15 Conditions used to explore whether chimpanzees understand the referential aspect of gaze (a) at target and (b) above target.

at-target trials. Thus, unlike three-year-old children, our apes behaved according to the predictions of a model that assumed that despite their excellent gaze-following abilities, they do not understand how gaze is related to subjective states of attention.

How, then, do we reconcile these findings with the earlier results concerning our animals' abilities to take into account whether someone's gaze is blocked by opaque barriers? Any such reconciliation must begin by abandoning Hume's argument by analogy and instead explore the possibility that identical behaviors may be generated or attended by different psychological representations. For example, in the case of their apparent understanding of how another's attention is blocked by opaque partitions, perhaps our original low-level model was too simplistic. Consider the following alternative explanation that accounts for their behavior without invoking an understanding of attention. Given that these animals possess a strong propensity to follow gaze, it seems quite plausible to suppose that this system is modulated by general learning mechanisms. Thus, with sufficient experience of following the "gaze" of others in the real world, primates may quickly learn how "gaze" interacts with objects and obstructions. In particular, they may simply learn that when they follow someone else's gaze to an opaque barrier, the space behind the barrier is no longer relevant. We have not yet tested this model against its alternatives. Instead, I use it to illustrate that even our apes' seemingly deep understanding of gaze on these opaque barrier tests deserves further empirical scrutiny.

BEYOND SEEING

It would be misleading to leave matters at this. After all, as surprised as we have been by what our apes appear to be telling us, if this were the sum total of our research on chimpanzees' understanding of mental states, it might only suggest that their species conceive of visual perception in a very different manner than our own. Although this would be interesting in its own right, other aspects of our research reveal a far more complicated (and intriguing) picture. Below, I briefly summarize just a few of these other projects. Almost without exception, our animals have consistently refused to be easily subsumed under the rubric of theory of mind.

Comprehending Pointing

The ability of chimpanzees to use their gestures to "indicate" or "choose" among people or things might lead one to wonder if this does not, by itself, indicate that they understand something about the mental state of attention. Indeed, we have even occasionally glossed our apes' actions as "pointing." Although we have done so with caution, it seems possible that they understand the referential significance of the gesture— that is, that they understand that gestures can be used to redirect attention toward particular locations or objects. We explored this possibility by teaching our adolescent chimpanzees (across dozens of trials) to pick a

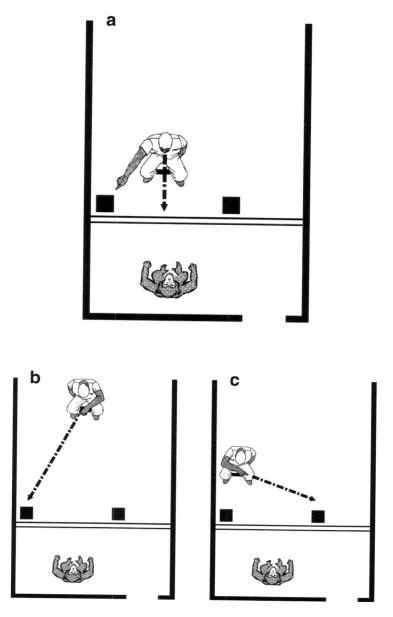

FIGURE 16   (a) With sufficient experience, chimpanzees learn to select a box to which the experimenter is pointing. (b) and (c) Alternative means of testing whether chimpanzees understand the referential significance of pointing or instead rely on the distance of the experimenter's hand from the correct box.

box to which their caregiver was pointing (Figure 16a; see Povinelli et al., 1997). Having trained them to do so, we considered several alternative ways in which they might be interpreting the task. On the one hand, they might have understood the idea of reference all along, but simply have needed some experience to apply it in this context. On the other hand, they might have learned a rule such as, "Look in the box closest to the caregiver's hand." We pitted these ideas against each other by confronting the apes, as well as very young children, with numerous configurations of someone pointing to two boxes. Figures 16b–c provide two of the most important contrasts. In one case, the experimenter's hand is equidistant from both boxes; in the other case, although the pointing gesture is referencing the correct box, the experimenter is physically closer to the incorrect box. In each of these cases (as well as several others), the apes performed according to the predictions of the distance-based model ("Choose the box closest to the experimenter's finger"). In contrast, 26-month-old children performed excellently even in the most difficult conditions (Figures 16a–c and 17a–c).

Distinguishing Intended from Unintended Actions

The distinction between intended and unintended (accidental or inadvertent) behavior is a core aspect of our human folk psychology and appears to be a distinction that is appealed to even in cultures very different from our own. Do chimpanzees interpret actions as being based on underlying intentions and hence distinguish between intended and unintended behavior? We investigated this by having our apes request juice or food from two strangers (see Povinelli et al., 1998). After several successful interactions with each person, we staged one of two events, each associated with one of the strangers. One of them accidentally spilled the juice intended for the ape, whereas the other slowly and deliberately poured the juice onto the floor, or, in another experiment, ate the food themselves. We wanted to know if, despite the fact that they failed to receive food from both of the strangers, the animals would distinguish between the two categories of action. When it came to the critical trials in which they had to choose one of the two strangers to give them more juice, the chimpanzees did not avoid the one who had intentionally wasted the juice. Young children, in contrast, begin to make such distinctions by about three years of age (e.g., Shultz et al., 1980; Yuill, 1984).

Cooperating with Others

Partially as a result of Crawford's early work in Robert Yerkes' laboratory, chimpanzees gained a reputation as being able to intentionally

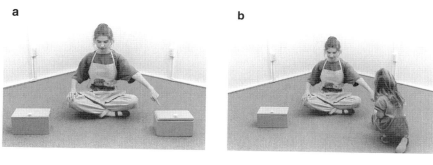

FIGURE 17   Two-year-old child responds to experimenter's pointing gesture correctly in (a) and (b) easy and (c) difficult conditions.

cooperate with each other. Crawford (1941) trained two apes to pull separate ropes to retrieve a box that was too heavy to manage alone. Although it is often overlooked, just to achieve this result required a large amount of training. Slowly, however, the animals began to attend to each other, and after about 50 sessions or so they began watching each other, eventually even exhibiting limited instances of touching each other when one wandered "off task." But was this evidence that they understood the intentional state of the partner or the mere conventionalization of behavior? Recently, O'Neill and I explored these possibilities. After training two "expert" apes to cooperate on a rope-pulling task, we then paired them (on several occasions), one by one, with familiar peers who were ignorant about the nature of the task (Povinelli and O'Neill, 2000). The results, perhaps more so than any others we have obtained, were truly captivating. There was not a single instance in which one of the expert apes attempted to direct the attention of the ignorant animals to the relevant features of the task. The expert apes were not oblivious to the need to have the other animal near the ropes, as they looked at the naive animal at levels far exceeding the frequency with which the naive animal looked at them. Indeed, there were a number of noteworthy occasions in which the naive partner approached and inspected the ropes, followed by the expert partner rushing forward to pull the rope he or she had typically used, while at the same time ignoring the other apes' actions. The expert partners appeared to realize that another animal needed to be near the ropes, but not that the other ape possessed his or her own subjective understanding of the situation.

REINTERPRETING BEHAVIOR

The traditional reaction to the remarkable behavioral similarity between humans and apes has been to conclude the presence of comparable psychological similarity. This was the solution counseled by Hume, and later adopted by Darwin and Romanes. In the remainder of this chapter, I sketch a different view, one more consistent with the mosaic of similarity and difference between humans and chimpanzees revealed by our research. But to appreciate this alternative view, we must reconsider the argument by analogy, and in particular its key assumption that introspection can successfully penetrate through the multiple levels of representation of the world formed by our neural circuitry, and retrieve the true causal connections between our thoughts and our actions.

Imagine our planet long before humans or the great apes had evolved. Imagine millions of generations of social primates evolving ever-more sophisticated brains to instantiate ever-more complex rules of social interactions, rules demanded by the complexities of group living. But, for the sake of argument, suppose that these rules did not appeal to mental states. Clever brains, in Humphrey's (1980) turn of phrase, but blank minds. Finally, suppose that one lineage, the human one, evolved a cognitive specialization linked to social understanding. Perhaps it evolved as a separate, domain-general faculty (in which case we might expect even more far-reaching differences between humans and chimpanzees than those discussed here); perhaps it evolved as a more specific system for social computations; finally, perhaps it evolved in conjunction with selection for a specialized language capacity. But however it arose, it seems likely that it did not replace the ancestral psychology of the social primates, but rather was woven into it. Like a tapestry into which new colors were added, new systems or subsystems may have been created while conserving an ancestral repertoire of social behaviors. Just as the morphological structures and systems of modern organisms reflect the baggage of developmental constraints laid down in the Cambrian, so too do psychological structures carry the ancient alongside the new.

This view essentially defeats Hume's hope that the spontaneous behavior of animals could serve as portals into their minds. Hume's error was twofold. First, he incorrectly assumed that an accurate introspective assessment of the mental states that *accompany* our own behaviors could yield an accurate inference as to the psychological cause of those behaviors. Furthermore, he compounded this error by advocating that we use the results of this introspective diagnosis to infer the mental states in other species by analogy. It is true that humans form internal representations of mental states such as desires, knowledge, and beliefs, and that these representations provide us with a useful means of anticipating what others will do. However, the basic blueprint of the behaviors with which we can respond may have evolved long before those higher-level representations were possible. Thus, the differences between humans and other species may stem first and foremost from underlying cognitive—not behavioral—specializations.

It may seem that I am advocating a form of dualism (such as epiphenomenalism) in which the representation of mental states accompany but play no causative role in behavior. On the contrary, I assume that humans evolved the capacity to represent the mental states of self and other because of their useful, causal connection to behavior. But until now we may have been thinking in the wrong way about this connection. As we have seen in the case of gaze following, two species may share a very similar behavior, controlled by similar low-level mechanisms, but interpret that behavior in very different ways. Does this mean that the human trait of representing the attentional aspect of gaze is superfluous? No, because although it may not be directly linked to the act of following gaze, the representation of attention (and other mental states) may play a profound role at a higher level of behavioral organization. Once humans evolved the capacity to represent other organisms not just as behaving beings, but as beings who attend, want, and think, it became necessary to use already-existing behaviors to respond to the world of the mind, as well as the world of behavior. This distinction may be especially important in those psychological arenas in which we seem to differ most from other species such as culture, pedagogy, and ethics. Determining the correct causal connections between individual behavioral elements and the mental states that ultimately control their expression may simply be a project beyond the reach of introspection alone (for a detailed critique of the argument by analogy, see Povinelli and Giambrone, in press).

More recently, we have begun to explore the differences that might exist between humans and chimpanzees in other psychological realms, such as understanding the physical world (see Povinelli, 2000). The results of this research suggest that here, too, chimpanzees are alert, cognizing beings—creatures that think and learn about the regularities of the interactions of physical objects in the world. And yet at the same time they appear to understand those interactions in ways that are very different from us. Unlike two-year- or three-year-old human children, chimpanzees do not appear to understand the causal structure that links one set of events to another. Just as they appear not to reason about unobservable mental states such as desires, perceptions, and beliefs, this new research suggests that they do not reason about unobservable physical causes such as gravity, force, or physical connection. Collectively, our work suggests the broader conclusion that chimpanzees may simply not have the ability to form ideas about phenomena that are so abstract that they are, in principle, unobservable. This may set direct limits on their ability to reason about virtually every aspect of the world around them.

My view, then, is that there may be profound psychological differences between humans and chimpanzees. Some will see this as a distinctly negative outlook, one that denies apes their rightful psychological standing alongside humans. But in the final analysis, evolution is about differences as well as similarities—indeed, were there no differences among species, there would have been no reason to postulate an evolutionary process in the first place. On this view, what could be a more constructive, more exciting project than to continue to explore the ways in which humans and chimpanzees have accommodated both the psychological structures of their mammalian ancestry and the innovations they continue to accumulate along their separate evolutionary journeys?

ACKNOWLEDGMENTS

This research was supported in part by National Science Foundation Young Investigator Award SBR-8458111. I thank James E. Reaux, Anthony Rideaux, and Donna Bierschwale for professional support in the testing of chimpanzees. I also thank Todd Preuss and Steve Giambrone for their generous effort in reviewing and editing the manuscript. Original artwork is by Donna Bierschwale. Photographs are by Donna Bierschwale and Corey Porché.

REFERENCES

Avis, J., and P. L. Harris. 1991. Belief-desire reasoning among Baka children: Evidence for a universal conception of mind. *Child Development* 62:460–467.
Baron-Cohen, S. 1994. How to build a baby that can read minds: Cognitive mechanisms in mindreading. *Current Psychology of Cognition* 13:513–552.

Blest, A. D. 1957. The function of eyespot patterns in the Lepidoptera. *Behaviour* 11: 209–255.

Burger, J., M. Gochfeld, and B. G. Murray, Jr. 1991. Role of a predator's eye size in risk perception by basking black iguana, *Ctenosaura similis*. *Animal Behaviour* 42:471–476.

Burghardt, G. M., and H. W. Greene. 1988. Predator simulation and duration of death feigning in neonate hognose snakes. *Animal Behaviour* 36:1842–1844.

Butterworth, G., and E. Cochran. 1980. Towards a mechanism of joint visual attention in human infancy. *International Journal of Behavioral Development* 3:253–272.

Butterworth, G., and N. Jarrett. 1991. What minds have in common is space: Spatial mechanisms serving joint visual attention in infancy. *British Journal of Developmental Psychology* 9:55–72.

Byrne, R. W., and A. Whiten. 1985. Tactical deception of familiar individuals in baboons (*Papio ursinus*). *Animal Behaviour* 33:669–673.

Call, J., B. A. Hare, and M. Tomasello. 1998. Chimpanzee gaze following in an object-choice task. *Animal Cognition* 1:89–99.

Corkum, V., and C. Moore. 1994. Development of joint visual attention in infants. Pps. 61–83. in C. Moore and P. Dunham, eds. *Joint Attention: Its Origins and Role in Development*. Hillsdale, N.J.: Lawrence Erlbaum.

Crawford, M. P. 1941. The cooperative solving by chimpanzees of problems requiring serial responses to color cues. *Journal of Social Psychology* 13:259–280.

Darwin, C. 1871. *The Descent of Man*. Reprinted. New York: Modern Library, 1982.

de Waal, F. B. M. 1989. *Peacemaking Among Primates*. Cambridge, Mass.: Harvard University Press.

Emery, N. J., E. N. Lorincz, D. I. Perrett, M. W. Oram, and C. I. Baker. 1997. Gaze following and joint attention in rhesus monkeys (*Macaca mulatta*). *Journal of Comparative Psychology* 111:286–293.

Flavell, J. H., S. G. Shipstead, and K. Croft. 1978. What young children think you see when their eyes are closed. *Cognition* 8:369–387.

Flavell, J. H., E. R. Flavell, F. L. Green, and S. A. Wilcox. 1980. Young children's knowledge about visual perception: Effect of observer's distance from target on perceptual clarity of target. *Developmental Psychology* 16:10–12.

Flavell, J. H., B. A. Everett, K. Croft, and E. R. Flavell. 1981. Young children's knowledge about visual perception: Further evidence for the level 1–level 2 distinction. *Developmental Psychology* 17:99–103.

Gallup, G. G., Jr., R. F. Nash, and A. L. Ellison, Jr. 1971. Tonic immobility as a reaction to predation: Artificial eyes as a fear stimulus for chickens. *Psychonomic Science* 23:79–80.

Gómez, J. C. 1990. The emergence of intentional communication as a problem-solving strategy in the gorilla. Pp. 333–355 in S. T. Parker and K. R. Gibson, eds. *Language and Intelligence in Monkeys and Apes: Comparative Developmental Perspectives*. Cambridge, U.K.: Cambridge University Press.

Goodall, J. 1986. *The Chimpanzees of Gombe: Patterns of Behavior*. Cambridge, Mass.: Belknap, Harvard University Press.

Gopnik, A., and P. Graf. 1988. Knowing how you know: Young children's ability to identify and remember the sources of their beliefs. *Child Development* 59:1366–1371.

Gopnik, A., A. N. Meltzoff, and J. Esterly. 1995. *Young Children's Understanding of Visual Perspective-Taking*. Poster presented at the First Annual Theory of Mind Conference, Eugene, Oregon.

Hains, S. M. J., and D. W. Muir. 1996. Effects of stimulus contingency in infant–adult interactions. *Infant Behavior and Development* 19:49–61.

Hume, D. 1739–1740/1911. *A Treatise of Human Nature*, 2 vols. London: Dent.

Humphrey, N. K. 1980. Nature's psychologists. Pp. 57–75 in B. D. Josephson and V. S. Ramachandran, eds. *Consciousness and the Physical World*. New York: Pergamon Press.

Karmiloff-Smith, A. 1992. *Beyond Modularity: A Developmental Perspective on Cognitive Science.* Cambridge, Mass.: MIT Press.

Köhler, W. 1927. *The Mentality of Apes,* 2nd ed. New York: Vintage Books.

Lempers, J. D., E. R. Flavell, and J. H. Flavell. 1977. The development in very young children of tacit knowledge concerning visual perception. *Genetic Psychology Monographs* 95:3–53.

Lillard, A. S. 1998a. Ethnopsychologies: Cultural variations in theory of mind. *Psychological Bulletin* 123:3–32.

Lillard, A. S. 1998b. Ethnopsychologies: Reply to Wellman (1998) and Gauvain (1998). *Psychological Bulletin* 123:43–46.

Masangkay, Z. S., K. A. McKluskey, C. W. McIntyre, J. Sims-Knight, B.E. Vaughn, and J. H. Flavell. 1974. The early development of inferences about the visual precepts of others. *Child Development* 45:357–366.

Moore, C. P. 1994. Intentionality and self–other equivalence in early mindreading: The eyes do not have it. *Current Psychology of Cognition* 13:661–668.

Moore, C., M. Angelopoulos, and P. Bennett. 1997. The role of movement and the development of joint visual attention. *Infant Behavior and Development* 20:83–92.

O'Neill, D. K., and A. Gopnik. 1991. Young children's ability to identify the sources of their beliefs. *Developmental Psychology* 27:390–397.

O'Neill, D. K., J. W. Astington, and J. H. Flavell. 1992. Young children's understanding of the role that sensory experiences play in knowledge acquisition. *Child Development* 63:474–490.

Perner, J., and J. Ogden. 1988. Knowledge for hunger: Children's problems with representation in imputing mental states. *Cognition* 29:47–61.

Perrett, D., M. Harries, A. Mistlin, J. Hietanen, P. Benson, R. Bevan, S. Thomas, M. Oram, J. Ortega, and K. Brierly. 1990. Social signals analyzed at the single cell level: Someone is looking at me, something touched me, something moved! *International Journal of Comparative Psychology* 4:25–55.

Povinelli, D. J. 1991. *Social Intelligence in Monkeys and Apes.* Ph.D. dissertation, Yale University, New Haven, Conn.

Povinelli, D. J. 1996. *Growing up ape.* Monographs of the Society for Research in Child Development 61(2, Serial No. 247):174–189.

Povinelli, D. J. 1999. Social understanding in chimpanzees: New evidence from a longitudinal approach. Pp. 195–225 in P. Zelazo, J. Astington, and D. Olson, eds. *Developing Theories of Intention: Social Understanding and Self-Control.* Hillsdale, N.J.: Erlbaum.

Povinelli, D. J. 2000. *Folk physics for apes: The chimpanzee's theory of how the world works.* Oxford: Oxford University Press.

Povinelli, D. J., and S. deBlois. 1992a. On (not) attributing mental states to monkeys: First, know thyself. *Behavioral and Brain Sciences* 15:164–166.

Povinelli, D. J., and S. deBlois. 1992b. Young children's (*Homo sapiens*) understanding of knowledge formation in themselves and others. *Journal of Comparative Psychology* 106:228–238.

Povinelli, D. J., and T. J. Eddy. 1994. The eyes as a window: What young chimpanzees see on the other side. *Current Psychology of Cognition* 13:695–705.

Povinelli, D. J., and T. J. Eddy. 1996a. *What young chimpanzees know about seeing.* Monographs of the Society for Research in Child Development 61(3, Serial No. 247):v–152.

Povinelli, D. J., and T. J. Eddy, 1996b. Chimpanzees: Joint visual attention. *Psychological Science* 7:129-135.

Povinelli, D. J., and T. J. Eddy. 1996c. Factors influencing young chimpanzees' (*Pan troglodytes*) recognition of attention. *Journal of Comparative Psychology* 110:336–345.

Povinelli, D. J., and T. J. Eddy. 1997. Specificity of gaze-following in young chimpanzees. *British Journal of Developmental Psychology* 15:213–222.

Povinelli, D. J., and S. Giambrone. 2000. Inferring other minds: Failure of the argument by analogy. *Philosophical Topics, 27*:161–201.

Povinelli, D. J., and L. R. Godfrey. 1993. The chimpanzee's mind: How noble in reason? How absent of ethics? Pp. 227–324 in M. Nitecki and D. Nitecki, eds. *Evolutionary Ethics.* Albany, N.Y.: SUNY Press.

Povinelli, D. J., and D. K. O'Neill. 2000. Do chimpanzees use their gestures to instruct each other? In S. Baron-Cohen, H. Tager-Flusberg, and D. J. Cohen, eds. *Understanding other Minds: Perspectives from Autism,* 2nd ed. Oxford, U.K.: Oxford University Press.

Povinelli, D. J., A. B. Rulf, K. R. Landau, and D. T. Bierschwale. 1993. Self-recognition in chimpanzees (*Pan troglodytes*): Distribution, ontogeny, and patterns of emergence. *Journal of Comparative Psychology* 107:347–372.

Povinelli, D. J., A. B. Rulf, and D. Bierschwale. 1994. Absence of knowledge attribution and self-recognition in young chimpanzees (*Pan troglodytes*). *Journal of Comparative Psychology* 180:74–80.

Povinelli, D. J., J. E. Reaux, D. T. Bierschwale, A. D. Allain, and B. B. Simon. 1997. Exploitation of pointing as a referential gesture in young children, but not adolescent chimpanzees. *Cognitive Development* 12:423–461.

Povinelli, D. J., H. K. Perilloux, J. E. Reaux, and D. T. Bierschwale. 1998. Young and juvenile chimpanzees' (*Pan troglodytes*) reactions to intentional versus accidental and inadvertent actions. *Behavioral Processes* 42:205–218.

Povinelli, D. J., D. T. Bierschwale, and C. G. Cech. 1999. Comprehension of seeing as a referential act in young children, but not juvenile chimpanzees. *British Journal of Developmental Psychology* 17:37–60.

Premack, D., and G. Woodruff. 1978. Does the chimpanzee have a theory of mind? *Behavioral and Brain Sciences* 1:515–526.

Reaux, J. E., L. A. Theall, and D. J. Povinelli. 1999. A longitudinal investigation of chimpanzees' understanding of visual perception. *Child Development* 70:275–290.

Redican, W. K. 1975. Facial expressions in nonhuman primates. Pp. 103–194 in L. A. Rosenblum, ed., *Primate Behavior,* Vol. 4. New York: Academic Press.

Ristau, C. A. 1991. Before mindreading: Attention, purposes and deception in birds? Pp. 209–222 in A. Whiten, ed. *Natural Theories of Mind: Evolution, Development and Simulation of Everyday Mindreading.* Cambridge, U.K.: Basil Blackwell.

Romanes, G. J. 1882. *Animal Intelligence.* London: Keagan Paul.

Ruffman, T. K., and D. R. Olson. 1989. Children's ascriptions of knowledge to others. *Developmental Psychology* 25:601–606.

Scaife, M., and J. Bruner. 1975. The capacity for joint visual attention in the infant. *Nature* 253:265–266.

Schaller, G. B. 1963. *The Mountain Gorilla: Ecology and Behavior.* Chicago: University of Chicago Press.

Shultz, T. R., D. Wells, and M. Sarda. 1980. Development of the ability to distinguish intended actions from mistakes, reflexes, and passive movements. *British Journal of Social and Clinical Psychology* 19:301–310.

Symons, L. A., S. M. J. Hains, and D. W. Muir. 1998. Look at me: Five-month-old infants' sensitivity to very small deviations in eye-gaze during social interactions. *Infant Behavior and Development* 21:531–536.

Theall, L. A., and D. J. Povinelli. 1999. Do chimpanzees tailor their attention-getting behaviors to fit the attentional states of others? *Animal Cognition* 2:207–214.

Tomasello, M. 1995. The power of culture: Evidence from apes. *Human Development* 38: 46–52.

Tomasello, M., J. Call, and B. Hare. 1998. Five primate species follow the visual gaze of conspecifics. *Animal Behaviour* 55:1063–1069.

Vinden, P. G., and J. W. Astington. 2000. Culture and understanding other minds. Pp. 503–519 in S. Baron-Cohen, H. Tager-Flusberg, and D. Cohen, eds. *Understanding other Minds: Perspectives from Autism and Cognitive Neuroscience.* Oxford, U.K.: Oxford University Press.

Wimmer, H., G. J. Hogrefe, and J. Perner. 1988. Children's understanding of informational access as a source of knowledge. *Child Development* 59:386–396.

Wolff, P. H. 1963. Observations on the early development of smiling. Pp. 113–138 in B. M. Foss, ed. *Determinants of Infant Behavior II.* London: Methuen & Co.

Yuill, N. 1984. Young children's coordination of motive and outcome in judgements of satisfaction and morality. *British Journal of Developmental Psychology* 2:73–81.

2

The Cognitive Neuroscience of Numeracy: Exploring the Cerebral Substrate, the Development, and the Pathologies of Number Sense

Stanislas Dehaene

INTRODUCTION

A basic unsolved problem in human cognition concerns how the brain computes meaning. As you read this text, what are the processes unfolding in your mind and brain that allow you to understand the message that I intend to communicate to you? What cerebral events enable you to recognize that the words "Einstein," "Eighteen," and "Elephant," in spite of their visual similarity, point to radically different meanings, which we can access in a fraction of a second? Recent studies in neuropsychology, brain imaging, and infant development have begun to provide a new perspective on this age-old problem. It is now well accepted that lesions to some specific areas of the human brain may cause highly selective semantic deficits (Caramazza, 1996; Hillis and Caramazza, 1995; McCarthy and Warrington, 1988, 1990). For instance, a patient may lose his or her knowledge of animals, but remains able to accurately define inanimate objects, food and vegetables, colors, numbers, or actions (Caramazza and Shelton, 1998). Brain-imaging studies have confirmed that processing of different categories of words such as proper names, animals, or tools is associated with distinct patterns of brain activation that may be similarly localized in different individuals (Damasio et al., 1996; Martin et al., 1995, 1996). Hence, specific cortical areas appear to participate in the representation of different categories of meaning.

Service Hospitalier Frédéric Joliot, INSERM, Orsay, France

In parallel, studies of human infants have revealed that they possess a much richer knowledge base than was initially suspected (Mehler and Dupoux, 1993; Pinker, 1995, 1997). Far from being a blank page, the human mind starts in life with a rich knowledge about objects, colors, numbers, faces, and language. This evidence supports the hypothesis that humans have been endowed by evolution with biologically determined predispositions to represent and acquire knowledge of specific domains (Barkow et al., 1992). Thus, if all languages distinguish between nouns and verbs, or between words for locations, colors, numbers, objects, animals, and persons, it may well be because the architecture of our brain is biologically constrained to focus learning on these domains, using partly dedicated cortical territories (Caramazza and Shelton, 1998).

If this framework has any validity, charting meaning in the human brain requires several steps: first, understanding the basic features specific to a given domain of knowledge; second, discovering whether and how some of these features have been extracted in the course of cerebral evolution and have been internalized in the brains of infants and animals; and third, identifying the neural circuits that underlie them and their internal organization. My research aims at fulfilling these goals in the specific case of arithmetic knowledge and at using this specific domain as a springboard for broader studies of the representation, development, and pathologies of semantic knowledge in the human brain.

THE COGNITIVE NEUROSCIENCE OF NUMERACY: PAST AND PRESENT EVIDENCE

A Biological Basis for Elementary Arithmetic?

In my research of the past decade, I have examined how a special category of words (numbers) are represented in the animal, infant, and adult brain. There are several reasons why numbers are excellent tools for psychological studies of word processing. First, number is a very circumscribed domain. There is only a small list of number words. Their lexicon, syntax, and semantics are clearly defined and have been formalized by mathematicians such as Peano or Russell. As David Marr (1982) remarked several years ago, understanding of any cognitive function requires an initial characterization of its computational demands. It is much easier to characterize what numbers mean than to understand, say, the semantics of verbs or of animal names.

A second advantage of studying numbers is that, for all their simplicity, they constitute one of the most important domains of human cognition. The number system is an indispensable tool for scientific activity. Arithmetic and mathematics in general are a unique achievement of our species, without which our technology-oriented society could never have developed. Philosophers have been pondering for thousands of years about the relations between mathematics and the human mind and brain. It is very exciting to think that, with the present experimental tools in cognitive neuroscience, we can begin to address this age-old question. It is also clear that this is not just a theoretical issue, but also one with considerable practical import. The loss of calculation and number processing abilities in brain-lesioned patients is a frequent and highly incapacitating deficit. So are developmental dyscalculia and mathematical disabilities in school children. The low level of mathematical competence, or downright innumeracy, particularly in Western countries relative to Asian countries, is a growing concern. By better understanding how mathematics is acquired and represented mentally and cerebrally, we may hope to improve education and rehabilitation methods in arithmetic.

A third advantage of studying the number domain is that there is already a rich database of experiments. I have reviewed many of these experiments in my recent book *The Number Sense* (Dehaene, 1997). In my opinion, they make the number domain one of the best documented instances of a biologically determined, category-specific domain of knowledge. (In that respect, number knowledge stands on a par with face processing, for which there is also considerable animal, infant, and adult lesion, brain-imaging, and even single-cell recording evidence for a biologically determined ability.) Strong evidence exists that animals and infants possess rudimentary arithmetic knowledge with a clear analogy and continuity with adult knowledge; that specific brain lesions can impair this number sense; and that a specific network of brain areas is active during number processing. This evidence provides support for my hypothesis that specific cerebral networks of the brain, through the course of a long evolutionary history, have internalized the numerical regularities of the world we live in, and it is this internalized representation of numerical quantities that enables us to understand number meaning. In this chapter, I review in turn the infant, animal, and adult behavioral and neurological evidence related to number processing. In each case, I outline what my own contributions to the domain are. Then I turn to the research projects I intend to pursue in the future.

Number Sense in Young Infants

There is now considerable evidence that infants possess some rudimentary knowledge of numbers in the first year of life, independently of language (Antell and Keating, 1983; Bijeljac-Babic et al., 1991; Koechlin et al., 1997; Moore et al., 1987; Simon et al., 1995; Sophian and Adams, 1987; Starkey and Cooper, 1980; Starkey et al., 1983, 1990; Strauss and Curtis,

1981; van Loosbroek and Smitsman, 1990; Wynn, 1992a, 1996; Xu and Carey, 1996). Discrimination of visual numerosity was first demonstrated in six-month- to seven-month-old infants using the classical method of habituation recovery of looking time (Starkey and Cooper, 1980). Following habituation to slides with a fixed number of dots, infants looked longer at slides with a novel number of objects. Discrimination was found for two versus three, and occasionally three versus four, but not four versus six. Recently, however, Xu and Spelke (2000) have established discrimination of 8 versus 16 dots (but not 8 versus 12), thus confirming that infants are not limited to small numbers.

In Starkey and Cooper's (1980) original experiment, dot density, spacing, and alignment were controlled for. In subsequent studies, the effect was replicated in newborns (Antell and Keating, 1983) and with various stimulus sets, including slides depicting sets of realistic objects of variable size, shape, and spatial layout (Strauss and Curtis, 1981), as well as dynamic computer displays of random geometrical shapes in motion with partial occlusion (van Loosbroek and Smitsman, 1990). Recently, the results have also been extended to the discrimination of visual events, such as a puppet making two or three jumps (Wynn, 1996). Using non-nutritive sucking techniques, it has also been established that four-day-old newborns distinguish two-syllable words from three-syllable words, even when phonemic content, duration, and speech rate are tightly controlled (Bijeljac-Babic et al., 1991). Thus, these results indicate that infants can perceive numerosity in various sensory modalities, independently of other more peripheral physical parameters. There are even experiments suggesting cross-modal numerosity matching in six-month- to eight-month-old infants (Starkey et al., 1983, 1990), although their replicability is disputed (Moore et al., 1987).

Infants use their numerosity perception abilities to predict and make sense of how objects behave in the external world. Wynn (1992a, 1995) presented five-month-old infants with real-life equivalents of the operations $1 + 1 = 2$ and $2 - 1 = 1$ (see Figure 1). To exemplify the $1 + 1 = 2$ operation, for instance, five-month-old infants were shown a toy being hidden behind a screen and then a second toy also being brought behind the same screen. To assess whether infants had developed the numerically appropriate expectation of two objects, their looking times were measured as the screen dropped and revealed either one, two, or three objects (objects being surreptitiously added or removed as needed). Infants looked systematically longer at the impossible outcomes of one or three objects than at the expected outcome of two objects. This suggests that they had internally computed an expectation of the outcome of two objects, although the two objects had never been presented together earlier.

Initial sequence: 1+1

1. First object is placed on stage   2. Screen comes up

3. Second object is added   4. Hand leaves empty

Possible outcome: 1+1=2

5. Screen drops...   revealing 2 objects

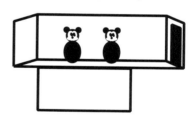

Impossible outcome: 1+1=1

5. Screen drops...   revealing 1 object

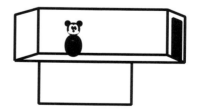

FIGURE 1 Design of infant experiments aimed at testing the understanding of a concrete operation "1 + 1 = 2" (redrawn from Wynn, 1992a).

In Wynn's original experiments, there were reasons to be skeptical about whether infants were really attending to number. The surprise reaction could also have been based on non-numerical parameters such as amount of stuff, object identity, or location. My own work with infants was directed at refuting such non-numerical interpretations (Koechlin et al., 1997). My colleagues and I replicated Wynn's experiments with two groups of infants. One group was tested in exactly the same conditions as Wynn's. In the other group, objects were placed on a rotating tray so that their location behind the screen was unpredictable. In both cases, infants still reacted to the numerically impossible events 1 + 1 = 1 and 2 – 1 = 2, thus indicating that they attended to a more abstract parameter than object location. A parallel experiment by Simon et al. (1995) showed that the identity of the objects could be surreptitiously modified on some trial, without altering the infants' surprise reaction. Together, these experiments suggest that infants encode the scenes they see using an abstract, implicit or explicit representation of the number of objects on the scene, irrespective of their exact identity and location.

Xu and Carey (1996) have examined which cues infants use to infer the presence of one or two objects. When infants under one year of age were presented with two highly discriminable objects alternatively popping out from the same screen, they did *not* expect to see two objects when the screen drops. Differences in object identity, by themselves, were not used by infants as a valid cue to number. Only if the trajectory followed by the objects was physically incompatible with it being a single object (as, for instance, if one object pops out from one screen on the right while the other object pops out from another spatially separate screen on the left) did preverbal infants infer the presence of two objects. This evidence is compatible with the hypothesis that numerosity is processed by a dedicated, informationally "encapsulated" module (Fodor, 1983) that takes as input information about object location and trajectory, but that is blind to object identity. Although the cerebral substrates of infants' numerical abilities are unknown, one may speculate that occipito-parietal pathways for spatial visual processing [the "where" system (Ungerleider and Mishkin, 1982)] may play a crucial role in numerosity extraction, thus explaining the special role of bilateral inferior parietal cortices in number processing in human adults (see below).

Number Sense in Animals

If neural networks of the infant brain are genetically biased toward elementary numerosity perception and arithmetic, one should find precursors of this ability in animals. Indeed, considerable evidence indicates that animals also possess numerosity discrimination, cross-modal numerosity perception, and elementary arithmetic abilities comparable to those of human infants (Boysen and Capaldi, 1993; Davis and Pérusse, 1988; Dehaene, 1997; Gallistel, 1989, 1990; Gallistel and Gelman, 1992). Like human infants, various animal species including rats, pigeons, raccoons, dolphins, parrots, monkeys, and chimpanzees have been shown to discriminate the numerosity of various sets, including simultaneously or sequentially presented visual objects as well as auditory sequences of sounds (Capaldi and Miller, 1988; Church and Meck, 1984; Mechner, 1958; Mechner and Guevrekian, 1962; Meck and Church, 1983; Mitchell et al. 1985; Pepperberg, 1987; Platt and Johnson, 1971; Rilling and McDiarmid, 1965). Cross-modal extraction of numerosity was observed, for instance, in rats (Church and Meck, 1984). Most such experiments included controls for non-numerical variables such as spacing, size, tempo, and duration of the stimuli (Church and Meck, 1984; Mechner and Guevrekian, 1962; Meck and Church, 1983).

Like human infants, animals also exhibit some abilities for elementary mental arithmetic. They can apply to their internal representations of number simple operations of approximate addition, subtraction and comparison (see Figure 2) (Boysen and Capaldi, 1993; Hauser et al., 1996; Rumbaugh et al., 1987; Woodruff and Premack, 1981). Wynn's $1 + 1 = 2$ and $2 - 1 = 1$ experiments with infants, for instance, have been replicated using a very similar violation-of-expectation paradigm with untrained monkeys tested in the wild (Hauser et al., 1996). In some experiments that do require training, there is excellent evidence that animals can generalize beyond the training range. A case in point is the remarkable recent experiment by Brannon and Terrace (1998, 2000). Monkeys were initially trained to press cards on a tactile screen in correct numerical order: first the card bearing one object, then the card bearing two, and so on up to four. Following this training, the monkeys were transferred to a novel block with numerosities five to nine. Although no differential reinforcement was provided, the monkeys readily generalized their smaller-to-larger ordering behavior to this new range of numbers. Such evidence suggests that a genuine understanding of numerosities and their relations can be found in monkeys.

There is also evidence that monkeys (Washburn and Rumbaugh, 1991), chimpanzees (Boysen and Berntson, 1996; Boysen and Capaldi, 1993; Matsuzawa, 1985), and even dolphins (Mitchell et al., 1985) and parrots (Pepperberg, 1987) can learn the use of abstract numerical symbols to refer to numerical quantities. Such symbolic number processing, however, requires years of training and is never found in the wild. Hence, it cannot be taken to indicate that exact symbolic or "linguistic" number processing is within animals' normal behavioral repertoire. It does indicate, however, that abstract, presumably nonsymbolic representations of number

FIGURE 2   A chimpanzee spontaneously selects the tray with the largest amount
of food, suggesting the availability of approximate addition and comparison operations (reprinted from Rumbaugh et al., 1987).

are available to animals and can, under exceptional circumstances,
be mapped onto arbitrary behaviors that can then serve as numerical
"symbols."

Changeux and I have developed a neural network model that repro-
duces the known features of animal numerical behavior and provides specific hypotheses about its neural basis (Dehaene and Changeux, 1993).

The network's input consists in a "retina" on which from one to five objects of various sizes can be presented. A first map encodes the locations of objects irrespective of their size and identity, an operation possibly performed by the occipito-parietal pathway. Following this "normalization" operation, each object is represented by an approximately equal number of active neurons, and hence their activity can be summed to yield an estimate of number. Number is therefore represented by the activity of an array of numerosity detector neurons, each responsive to a specific range of input numerosities. Interestingly, number-sensitive neurons similar to those predicted by our model have been recorded by Thompson et al. (1970) in the associative cortex of cats.

The Phylogenetic Continuity of Number Sense

To demonstrate that human abilities for arithmetic have a biological basis with a long evolutionary history, it is not sufficient to demonstrate that animals possess rudimentary number processing abilities. One must also show that this is not a case of parallel evolution, and that there are deep relations between human and animal abilities that suggest a phylogenetic continuity. Indeed, a considerable part of my chronometric work with human adults has been dedicated to documenting two characteristics of number processing that are shared by humans and animals: the distance effect and the size effect (see Figure 3). The distance effect is a systematic, monotonous decrease in numerosity discrimination performance as the numerical distance between the numbers decreases. The size effect indicates that for equal numerical distance, performance also decreases with increasing number size. Both effects indicate that the discrimination of numerosity, like that of many other physical parameters, obeys Fechner's law.

It is perhaps unsurprising that humans, like other animals, suffer from distance and size effects when processing the numerosity of sets of objects (Buckley and Gillman, 1974; van Oeffelen and Vos, 1982). More interestingly, these effects still hold when human subjects process symbolic stimuli such as Arabic numerals. The time and error rate for humans to compare two numerals increases smoothly, both with number size and with the inverse of the numerical distance between the two numbers. Thus, it is harder to decide that 1 is smaller than 2 than to reach the same decision with 1 and 9; and it is harder still with 8 and 9 (Buckley and Gillman, 1974; Moyer and Landauer, 1967). My own work has helped specify the mathematical equations for comparison time (Dehaene, 1989). I have also explored the interesting case of two-digit number comparison (Dehaene et al., 1990; Hinrichs et al., 1981). One might suppose that, in order to compare 2 two-digit numbers, subjects first compare the leftmost

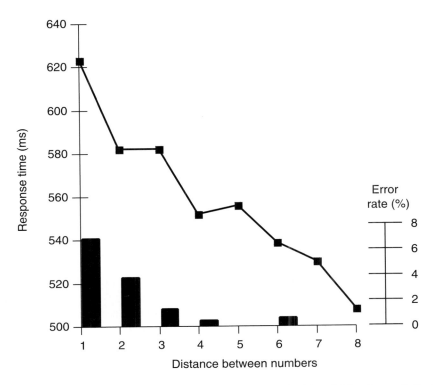

FIGURE 3   The distance effect. The time needed to decide which of two digits is larger, as well as the error rate, decreases as the distance between the two numerals gets larger (redrawn from data in Moyer and Landauer, 1967).

digits and then, if necessary, the rightmost digits. Experiments show, however, that comparison time is still a continuous function of the distance between the two numbers; that the rightmost digit has an effect even when the leftmost digit suffices to respond; and that there is little or no discontinuity in reaction time at decade boundaries. Hence, two-digit numbers seem to be compared as a whole quantity and in an analog fashion.

In general, number comparison data are best explained by supposing that Arabic numerals are immediately coded mentally in an analogical representation of quantities similar to the one found in animals. My theory, which is validated by a considerable number of experimental results, proposes that numbers are not compared as digital symbols, but as analog quantities that can be represented by distributions of activation on an internal continuum, a "number line" obeying Fechner's law (Dehaene et al., 1990; Restle, 1970). The numerical distance and size effect are a characteristic signature of this mode of representing numbers. The fact that they are found in all sorts of number processing tasks, whether in animals or in humans, with numbers conveyed by symbols or by the numerosity of sets of objects, is crucial to my hypothesis that a "number sense" is part of our evolutionary heritage.

As a matter of fact, distance and size effects have been observed, not only during number comparison, but also during calculation (Ashcraft, 1992), addition or multiplication verification (Ashcraft and Stazyk, 1981), parity judgments (Dehaene et al., 1993), or number reading (Brysbaert, 1995). Interestingly, they are also found when subjects decide whether two digits are the same or different (Duncan and McFarland, 1980). I have shown that, even when subjects try to focus only on the physical differences between two digits, they are slower to respond "different" to two numerically close numbers such as 8 and 9 than to two distant numbers such as 2 and 9 (Dehaene and Akhavein, 1995). Independent evidence suggests that, every time a number is seen, a representation of the corresponding quantity is accessed automatically and irrepressibly (Brysbaert, 1995; Henik and Tzelgov, 1982; LeFevre et al., 1988; Tzelgov et al., 1992; Zbrodoff and Logan, 1986). In my future work, I plan to image this conscious and unconscious access to number semantics using a masked priming situation I have developed (Dehaene et al. 1998c).

Two studies with normal subjects have also provided further proof of the ubiquity of the quantity representation in human number processing. First, Mehler and I have discovered that in all languages, the frequency with which numerals are used in written or spoken texts decreases exponentially with number size, with only local increases for round numbers (Dehaene and Mehler, 1992). This effect can be entirely explained by supposing that language-specific number words are used to access a language-independent quantity representation, a universal compressive (Fechnerian) number line.

Second, a reaction time effect, the spatial-numerical association of response codes (SNARC), indicates that the number line is intimately associated with representations of space (Dehaene et al., 1993). The SNARC effect is the fact that whenever subjects do a bimanual response to numbers they respond faster with the right hand to larger numbers and with the left hand to smaller numbers. My own work used parity judgments, but others have replicated the SNARC effect with symmetry judgments (Huha et al., 1995) or rhyme judgments (Fias et al., 1996). The results indicate that the effect does not depend on the absolute size of the numbers, but only on their relative size. Its direction—small to the left and large to the right—does not change in left-handers, when the hands are crossed, or with mirror-imaged digits or number words. However, it does reverse in Iranian students who recently immigrated to France and who write from right to left. Thus, there seem to be strong ties between numbers and space, but the direction of this spatial-numerical association is influenced by culture. In my book, *The Number Sense* (Dehaene, 1997), I speculate that number forms—the vivid, often colorful spatial patterns that some synesthetic subjects claim to experience when thinking about numbers (Galton, 1880; Seron et al., 1992)—are an unusually enriched version of this spatial-numerical association that exceptionally reaches access to consciousness.

Losing Number Sense

One final piece of evidence is required to demonstrate that the elementary understanding and manipulation of numerical quantities is part of our biological evolutionary heritage. One should show that it has a dedicated neural substrate. Animals and humans should share a set of homologous brain areas whose networks are dedicated to representing and acquiring knowledge of numerical quantities and their relations. This hypothesis has guided my neuropsychological and brain-imaging research for almost a decade. Although the demonstration is still far from complete, we now have two strong arguments to suggest that number processing is associated with a specific cerebral network located in the inferior intraparietal area of both hemispheres. First, neuropsychological studies of human patients with brain lesions indicate that the internal representation of quantities can be selectively impaired by lesions to that area (Benton, 1987; Cipolotti et al., 1991; Dehaene and Cohen, 1997; Gerstmann, 1940; Takayama et al., 1994; Warrington, 1982). Second, brain-imaging studies reveal that this region is specifically activated during various number processing tasks (Dehaene, 1996; Dehaene et al., 1996; Kiefer and Dehaene, 1997; Roland and Friberg, 1985; Rueckert et al., 1996).

It has been known for at least 80 years that lesions of the inferior parietal region of the dominant hemisphere can cause number processing deficits (Gerstmann, 1940; Hécaen et al., 1961; Henschen, 1920) (see Figure 4). In some cases, comprehending, producing, and calculating with numbers are globally impaired (Cipolotti et al., 1991). In others, however, the deficit may be selective for calculation and spare reading, writing, spoken recognition, and production of Arabic digits and numerals (Dehaene and Cohen, 1997; Hécaen et al., 1961; Takayama et al., 1994; Warrington, 1982).

My colleague Laurent Cohen and I have recently suggested that the core deficit in left parietal acalculia is a disorganization of an abstract semantic representation of numerical quantities rather than of calculation processes per se (Dehaene and Cohen, 1995; 1997). One of our patients, Mr. Mar (Dehaene and Cohen, 1997), experienced severe difficulties in calculation, especially with single-digit subtraction (75 percent errors). He

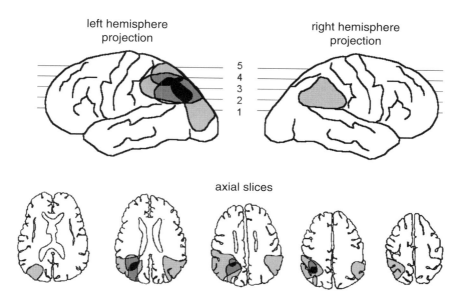

FIGURE 4   Approximate localization of the lesions in patients suffering from acalculia in the context of Gerstmann's syndrome. Although these patients may still read and write numbers, they exhibit a selective deficit of the comprehension and mental manipulation of numerical quantities (redrawn from Dehaene et al., 1998b).

failed on problems as simple as 3 − 1, with the comment that he no longer knew what the operation meant. His failure was not tied to a specific modality of input or output, because the problems were simultaneously presented visually and read out loud and because he failed in both overt production and covert multiple-choice tests. Moreover, he also erred on tasks outside of calculation per se, such as deciding which of two numbers is the larger (16 percent errors) or what number falls in the middle of two others (bisection task: 77 percent errors). He easily performed analogous comparison and bisection tasks in non-numerical domains such as days of the week, months, or the alphabet (What is between Tuesday and Thursday? February and April? B and D?), indicating that he suffered from a category-specific deficit for numbers. We have now made several observations of patients with dominant-hemisphere inferior parietal lesions and Gerstmann's syndrome. All of them showed severe impairments in subtraction and number bisection, suggesting disturbance to the central representation of quantities.

A "developmental Gerstmann syndrome," also called developmental dyscalculia, has been reported in children (Benson and Geschwind, 1970; Spellacy and Peter, 1978; Temple, 1989, 1991). Some children show a highly selective deficit for number processing in the face of normal intelligence, normal language acquisition, and standard education. Paul (Temple, 1989), for instance, is a young boy who suffers no known neurological disease, has a normal command of language, and uses an extended vocabulary, but has experienced exceptionally severe difficulties in arithmetic since kindergarten. At the age of 11, he remained unable to multiply, subtract, or divide numbers and could only add some pairs of digits through finger counting. Paul can easily read nonwords as well as infrequent and irregular words such as *colonel*. However, he makes word substitution errors only when reading numerals, for instance, reading *one* as "nine" and *four* as "two."

Although there is a dearth of accurate brain-imaging data on such developmental dyscalculia cases, it is tempting to view them as resulting from early damage to inferior parietal cortices that hold an innate, biologically determined representation of numbers. Part of my research project includes exploring the neural substrates of number processing in infants and young children with modern brain-imaging techniques and studying whether these networks are impaired in dyscalculia cases. With two American colleagues, Streissguth and Kopera-Frye, we have shown that children that have been exposed to alcohol in utero suffer from specific deficits of number sense (Kopera-Frye et al., 1996), suggesting that the laying down of numerical networks begins long before birth. Fetal alcohol syndrome (FAS) infants may thus provide a useful group for imaging early impairments of number processing. Developmental dyscalculia can be a devastatingly incapacitating deficit. I have the hope that a better understanding of the neural networks of number sense, and how they dysfunction when lesioned, may lead to better and earlier rehabilitation programs.

Although the inferior parietal region seems to play a crucial role in number sense, it is important to note that it is not the only brain region involved in number processing in adults. The phrenological notion that a single area can hold all the knowledge about an entire domain such as arithmetic has to give way to a more parallel view of number processing in the brain. Multiple brain areas are involved, whether for identifying Arabic numerals, writing them down, understanding spoken number words, retrieving multiplication facts from memory, or organizing a sequence of multidigit calculations. To explore these circuits, Cohen and I have studied tens of brain-lesioned patients with number processing deficits, and we have demonstrated the existence of previously unknown dissociations (Cohen and Dehaene, 1991, 1994, 1995, 1996; Cohen et al., 1994; Dehaene and Cohen, 1991, 1994, 1995, 1997):

- Split-brain patients: We have now seen three patients with posterior lesions restricted to the corpus callosum and whose posterior hemispheres were consequently disconnected, enabling us to study the numerical functions of the left and right hemispheres (Cohen and Dehaene, 1996). The results indicated that both hemispheres can recognize Arabic digits, convert them to quantities, and compare them. Only the left hemisphere, however, is able to name numerals and to perform exact calculations. Our results have confirmed and extended previous results obtained with surgical cases of callosal lesions (Gazzaniga and Hillyard, 1971; Gazzaniga and Smylie, 1984; Seymour et al., 1994).

- Large left-hemispheric lesions: Two single cases of patients with large lesions of the posterior left hemisphere have enabled us to confirm that the right hemisphere may not be devoid of numerical abilities. One patient was able to provide approximate answers to calculations he could not perform in exact form (Dehaene and Cohen, 1991). The other, despite severe acalculia and number-naming deficits, remained able to access encyclopedic knowledge of quantities, dates, and other familiar numbers (Cohen et al., 1994).

- Pure alexia: In two cases of patients with pure alexia, the visual input to the left hemisphere was destroyed by a left inferior occipito-temporal lesion. Yet again, we were able to confirm the presence of intact Arabic digit identification and comparison abilities, presumably supported by right-hemispheric circuits (Cohen and Dehaene, 1995).

- Neglect dyslexia for number: One single case of a patient with a peculiar Arabic number reading deficit enabled us to study the cognitive mechanisms underlying number reading (Cohen and Dehaene, 1991).

- Subcortical acalculia: One patient, a retired school teacher, with a lesion to the left lenticular nucleus failed to recite the multiplication table, together with prayers, poetry, nursery rhymes, or even the alphabet (Dehaene and Cohen, 1997). Number sense was preserved, however, as indicated by relatively preserved comparison, subtraction, and interval bisection abilities. Hence, there was a double dissociation with the deficits observed in parietal acalculia and Gerstmann's syndrome. We suggested that a left cortico–subcortical circuit, connected to perisylvian language areas, contributes to the rote retrieval of arithmetic tables together with other verbal and nonverbal routine operations.

Many additional acalculia cases have been reported by other groups (Campbell and Clark, 1988; Caramazza and McCloskey, 1987; Cipolotti

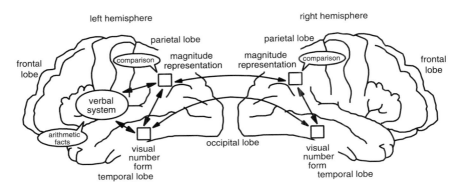

FIGURE 5   Schematic diagram of the main brain areas thought to be involved in simple number processing and their functional interactions (redrawn from Dehaene and Cohen, 1995).

and Butterworth, 1995; Cipolotti et al., 1994, 1995; Dagenbach and McCloskey, 1992; Deloche and Seron 1982a, 1982b, 1984, 1987; Macaruso et al., 1993; McCloskey, 1992, 1993; McCloskey and Caramazza, 1987; McCloskey et al., 1985, 1986, 1991a, 1991b, 1992; Noel and Seron, 1993; Seron and Deloche, 1983; Sokol et al., 1989, 1991; Warrington, 1982). Based on them, Cohen and I have proposed a tentative model of the cerebral circuits implicated in calculation and number processing: the *triple-code model* (Dehaene, 1992; Dehaene and Cohen, 1995) (see Figure 5). This model helps us make sense of the numerical deficits that patients exhibit consecutive to a focal brain lesion. It also does a fair job of accounting for chronometric data in normal subjects.

Functionally, the model rests on three fundamental hypotheses. First, numerical information can be manipulated mentally in three formats: an analogical representation of quantities, in which numbers are represented as distributions of activation on the number line; a verbal format, in which numbers are represented as strings of words (e.g., thirty seven); and a visual Arabic number form representation, in which numbers are represented as a string of digits (e.g., 37). Second, transcoding procedures enable information to be translated directly from one code to the other. Third, each calculation procedure rests on a fixed set of input and output codes. For instance, we have seen how number comparison works on numbers coded as quantities on the number line. Likewise, the model postulates that multiplication tables are memorized as verbal associations between numbers represented as a string of words and that multidigit operations are performed mentally using the visual Arabic code.

Anatomically, our neuropsychological observations have enabled us to flesh out the model and to associate tentative anatomical circuits to each function. We speculate that the mesial occipito-temporal sectors of both hemispheres are involved in the visual Arabic number form; that the left perisylvian areas are implicated in the verbal representations of numbers (as with any other string of words); and, most crucially, that the inferior parietal areas of both hemispheres are involved in the analogical quantity representation. Note that the redundant representation of the visual and quantity codes in the left and right hemispheres can explain why pure alexia patients (Cohen and Dehaene, 1995) or patients with major left-hemispheric lesions (Cohen et al., 1994; Dehaene and Cohen, 1991; Grafman et al., 1989) remain able to compare two Arabic numerals. It also explains that in callosal patients, number comparison remains feasible by both hemispheres (Cohen and Dehaene, 1996; Seymour et al., 1994). However, the unilateral, left-hemispheric lateralization of the verbal code and of the calculation abilities that depend on it explains why a left-hemispheric lesion suffices to abolish these abilities and why the right hemisphere of callosal patients cannot read numbers aloud or calculate with them.

Imaging Number Sense in the Living Human Brain

Functional brain-imaging techniques now provide new tests of the organization of cognitive processes. In itself, information about cerebral localization is not particularly interesting to cognitive scientists. Once a specific brain area has been localized, however, it becomes possible to ask what are the parameters that make it more or less active; what tasks it is responsive to; and what aspects of the stimulus, task, or response do *not* affect its state of activity. With dynamic methods such as electroencephalography or magnetoencephalography, it is also possible to record how quickly and for how long a region activates. In all these respects, the triple-code model makes specific predictions. Most critically, it predicts that the left and right inferior parietal areas should be active during various quantitative number processing tasks, and that their activation should depend purely on quantitative parameters such as number size and numerical distance, but not on the input or output modality or on the notation used for numbers. I have tested these predictions using positron emission tomography (PET) (Dehaene et al., 1996), electroencephalography (event-related potentials or ERPs; Dehaene, 1996; Dehaene et al., 1994; Kiefer and Dehaene, 1997); and, more recently, functional magnetic resonance imaging (fMRI) (Chochon et al., 1999; Dehaene et al., 1999).

Roland and Friberg (1985) were the first to monitor blood flow changes during calculation as opposed to rest. When subjects repeatedly subtracted 3 from a given number, activation increased bilaterally in the inferior parietal and prefrontal cortex. These locations were later confirmed using fMRI (Burbaud et al., 1995; Rueckert et al., 1996). Because the serial subtraction task imposes a heavy load on working memory, it was not clear which of these activations were specifically related to number processing. However, my own experiments using simpler tasks have confirmed that the inferior parietal area is specifically involved in number processing. In a PET study of multiplication and comparison of digit pairs, my colleagues and I again found bilateral parietal activation confined to the intraparietal region (Dehaene et al., 1996). This confirmed results obtained with a coarser resolution electroencephalogram method (Inouye et al., 1993) as well as with single-unit recordings in neurological patients (Abdullaev and Melnichuk, 1996).

In more detailed studies, I have begun to examine the prediction that inferior parietal cortex activation reflects the operation of an abstract quantity system largely independent of input and output modalities as well as of the specific arithmetic task involved. In one study, I used high-density recordings of ERPs during a number comparison task to study the cerebral basis of the distance effect. An additive-factor design was used in which three factors were varied: number notation (Arabic or verbal), numerical distance (close or far pairs), and response hand (right or left). The results revealed that inferior parietal activity was modulated by the numerical distance separating the numbers to be compared, but not by the notation used to present them (Dehaene et al., 1996). Notation did have a significant effect around 150 ms after the visual presentation of the stimuli, indicating bilateral processing for Arabic digits, but unilateral left-hemispheric processing for visual number words (in agreement with the triple-code model). By about 200 ms, however, ERPs were dominated by a parietal distance effect with a significant lateralization to the right, and without further influence of notation. The response hand effect, which emerged as early as 250 ms after the stimulus, provided an upper bound on the duration of the numerical comparison process.

A similar study of number multiplication showed that inferior parietal activity lasts longer during multiplication of two large digits than during multiplication of two small digits, again regardless of the modality of presentation of the operands (auditory or visual) (Kiefer and Dehaene, 1997). The main difference with my previous study was that the ERP effects, though always bilateral, were stronger over the *left* inferior parietal area during multiplication, but stronger over the *right* parietal area during number comparison. We recently replicated this modulation by task demands using fMRI (Chochon et al., 1999). Relative to letter reading, digit comparison yielded greater activity in the right inferior parietal area, multiplication greater activity in the left, and subtraction a bilateral increase (Figure 6). In spite of these small variations, however, an exten-

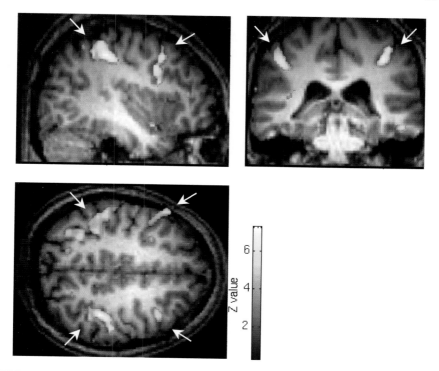

FIGURE 6 Bilateral activation of the intraparietal sulcus, postcentral sulcus, and prefrontal areas during single-digit subtraction in a single subject (data from Chochon et al., 1999).

sive bilateral intraparietal network was common to all three tasks. In agreement with neuropsychological observations of inferior parietal acaculia, I have suggested that this network represents the cerebral basis of the number sense.

FRONTIERS IN THE COGNITIVE NEUROSCIENCE OF NUMERACY

Outline of Future Research Projects

In the next decade, I plan to continue exploring the cerebral bases of human cognitive functions using number processing as a prime example. My research projects are headed in two main directions.

First, I plan to further test and validate the Dehaene–Cohen triple-code model of number processing. Despite the considerable progress that has been made in understanding the cerebral bases of numeracy in the past decade, the present state of our knowledge cannot be considered definitive. Most importantly, although we are beginning to understand the organization of the adult cerebral networks for number processing—and further experiments are needed in that domain too—we still have little or no information as to how these networks develop. Dehaene-Lambertz and I are planning to run brain-imaging studies to determine if, when, and how inferior parietal circuits are involved in number processing during infancy and childhood. Our experiments, based on ERP recordings as well as on the newer technique of near-infrared spectroscopy (NIRS), should help us image the normal development of number sense and its disorganization in dyscalculia. We also have a very poor understanding of how the posterior cerebral networks for verbal, Arabic, and quantitative number processing are coordinated during the execution of complex sequential operations. Imaging will help us understand the anatomical organization of numerical representations, particularly in the intraparietal sulcus, and their relations to prefrontal cortices.

Second, the detailed knowledge that we have accrued in the numerical domain will be used to bring new light on more general issues in cognitive neuroscience. I shall use my expertise in the number domain to study the differences between conscious and unconscious processing of numerical information. My students and I have developed a masked priming task in which we have clear behavioral evidence of unconscious access to number semantics. We plan to use brain-imaging techniques to image the processing of unconscious masked primes and thereby to throw some light on what distinguishes conscious and unconscious processing in the human brain. Part of my research will also assess the existence of category-specific representations of word meaning. I shall examine whether other categories of meaning than numbers, such as body parts, animals, foods, and so on, can yield to a similar analysis using convergent behavioral, neuropsychological, and neuroimaging methods.

Research Setting

My research is conducted in the Service Hospitalier Frédéric Joliot (SHFJ) in Orsay, about 15 miles south of Paris. The center is equipped with two PET scanners, one cyclotron, two fMRI scanners, equipment for high-density recordings of ERPs from 128 electrodes, and a major computer network. SHFJ is run by the French Atomic Energy Commission, which provides excellent technical support for developments in imaging and computer science. Within it, my team specializes on the cognitive neuroscience of higher-level functions of the human brain.

My team also benefits from collaborations with two other colleagues.

First, Dehaene-Lambertz, whose research focuses on early language development and infant brain-imaging studies, collaborates on studies of numerical development. She is running 64-channel ERP studies of infants and young children with her own equipment set up in a neuropediatrics unit near Paris, with good recruitment of normal, brain-lesioned, and retarded infants and children. Second, my colleague, Laurent Cohen, a neurologist and an M.D./Ph.D., will continue in our longstanding collaboration on neuropsychology studies of patients with language and/or calculation impairments. Dr. Cohen does clinical work and research at the Hopital de la Salpêtrière in Paris, a well-known hospital with a specialization in neurology and neuropsychological dating back to the foundational studies of Broca, Charchot, or Pinel. Patients recruited at Salpêtrière will be tested behaviorally and using fMRI and ERP techniques, either at SHFJ or using Salpêtrière's own resources (fMRI and whole-head magneto-encephalography).

Brain-imaging research is costly and fast changing. The James S. McDonnell Centennial Fellowship will considerably help to achieve the brain-imaging and the developmental parts of the research project. In the near future, Dehaene-Lambertz and I will explore the suitability of novel equipment for NIRS for brain-imaging studies of infant cognition. This novel imaging technique is sensitive to brain activation like fMRI, but it is much lighter and less invasive so that it can be applied readily to infants and young children. With it, I hope to image the development of the anatomical networks of number sense since infancy.

Imaging the Development of Number Sense

One of the most important gaps in our current understanding of number sense is the absence of any data concerning the cerebral underpinnings of infant numerical abilities and of children's mathematical development. In *The Number Sense* (Dehaene, 1997), I speculated that infant and animal abilities for number processing indicate an early functioning of the cerebral networks of number sense in the inferior parietal lobe. I also suggested that acquisition of number words and of the counting sequence is accompanied by the development of systematic connections between the quantity representation in the parietal lobe and the rest of the language system. Dehaene-Lambertz and I will directly put these hypotheses to a test.

Event-Related Potential Studies in Infants

In the very near future, we will start running studies of infant number processing using high-density recordings of ERPs from 64 channels, a technique with which we now have considerable experience (e.g., Dehaene-Lambertz and Dehaene, 1994). We shall start by replicating numerosity discrimination studies in the auditory modality, and later in the visual modality. ERPs will be recorded as infants are habituated to repeated sequences comprising an identical number of sounds (e.g., two), but with varied tempo, spacing, and identity of the individual sounds. Occasionally, sequences comprising a novel number of sounds (e.g., three) will be presented. ERPs will be monitored for the presence of a response to numerical novelty. A similar design, again controlling for all non-numerical parameters, will be used in the visual modality. A preliminary pilot study with adults indicates that we may find an auditory mismatch negativity similar to that found in other auditory mismatch situations (Näätänen, 1990), but also possibly an amodal parietal difference common to auditory and visual stimuli, and hence plausibly related to numerosity processing.

If this infant study works in a similar way, we also plan to replicate Wynn's infant addition and subtraction paradigms using ERPs, looking for cerebral correlates of infant's surprise response to an unexpected numerosity. In older children, between two and four years of age, we also hope to run ERP studies of how the child's growing vocabulary of number words connects to number meaning. Wynn (1992b) has identified a stage in numerical development during which children can recite number words and yet fail to relate them to specific quantities (for instance they can count "one, two, three," but fail to relate the final word "three" to the cardinal of the counted set; and they cannot point to a card with a specific number of objects, such as two). Imaging children listening to number words before and after that stage should enable us to test the hypothesis that the crucial change, at that stage, is the connection of the left-hemispheric verbal system to the nonverbal bilateral parietal quantity system.

Near-Infrared Spectroscopy

In the past four years, NIRS has been used in cerebral functional activation studies to monitor changes of cerebral oxygenation during sensory, motor, or cognitive tasks in human adults (Gratton et al., 1995; Kleinschmidt et al., 1996; Meek et al., 1995), including mathematical problem solving (Hoshi and Tamura, 1997). Briefly, laser light is shone through the scalp at several wavelengths, and scattered light is then detected through photodiodes only a few centimeters away. The intensity of light scattered at specific wavelengths is a direct reflection of oxyhemoglobin and deoxyhemoglobin concentration in the underlying tissue. Hence, NIRS is sensitive to the very same parameters that fMRI measures (brain oxygen-level dependent contrast). The equipment is so much lighter, however, as to be portable. Furthermore, the cost is low, and the signals can be sampled at up to 50 Hz, allowing for high temporal resolution studies. Ongoing research by Gratton and Fabiani at the University of Missouri suggests that the technique might also be directly sensitive to neuronal activity rather than to subsequent hemodynamic changes, allowing for an even greater temporal resolution. The only serious disadvantage is that, currently, a single emittor-photodiode pair can only detect activation in a single, relatively large brain volume of a few cubic centimeters.

There are several reasons why NIRS may become the ideal technique for brain-imaging studies in infants and in young children. First, fMRI and PET studies of young subjects are ethically questionable and impractical because of the requirements to suppress head motion and remain still but cognitively active. Second, ERP studies have an excellent temporal resolution but are notoriously poor at reconstructing the localization of brain activity. NIRS seems to be the only technique to date that can provide unquestionable functional brain localization information in very young subjects. Third, infants have a small translucent skull and small jaw muscles that should permit much cleaner signal acquisition than in adults. Fourth, NIRS is applicable only to a depth of 1–2 cm, a severe limitation for adult heads but one that should not prevent imaging the entire depth of the cortex in infants.

The Centennial Fellowship will help evaluate the feasibility of NIRS studies of infant and child development in the domains of language and arithmetic. Following pilot studies with simple auditory and visual stimulation, I hope to be able to run studies similar to the above ERP studies of numerosity discrimination in infants using NIRS. In the near future, only a technique such as NIRS seems to have the potential to directly test the hypothesis that the inferior parietal lobe is active during number processing as early as during infancy. But, as novel imaging techniques constantly emerge, I intend to use part of the grant money to follow all new developments in brain imaging that would be suited to noninvasive cognitive brain-imaging studies in young children.

Imaging the Anatomy of Number Sense as a Function of Expertise

The impact of education on the cerebral networks of number sense can also be tested in more classical studies using fMRI and ERPs in adults. My colleagues and I have already performed fMRI and ERP studies in which two operations—single-digit multiplication, which is learned through rote verbal memorization; and subtraction, which is performed through nonmemorized quantity manipulations—correlate with strikingly distinct cerebral activation (Chochon et al., 1999). In another very recent experiment, we showed that exact and approximate calculations also rely on strikingly distinct circuits (Dehaene et al., 1999). I envision several extensions of this work. First, we can explicitly train adult subjects on various number facts using different strategies, and we can image the concommittent changes in brain activation. Second, we can image activation in the same basic arithmetic tasks in populations with various levels of expertise such as literary versus mathematically oriented students, or even possibly adults with prodigious calculation abilities as compared with normal adults. All of these studies should help us understand how the cerebral architecture for number processing changes with education and expertise. Third, a fascinating experiment would involve imaging cerebral activity during calculation and number processing in adults with or without number forms (Galton, 1880), thus addressing directly whether synesthesia for numbers corresponds to a genuine enrichment of the cerebral map for numbers.

Fourth, and perhaps most importantly, we may also image the cerebral correlates of specific deficits of number sense, either by comparing normal children with children with specific retardation in arithmetic (dyscalculia) or by studying adult brain-lesioned patients before and after recovery of calculation abilities. Once a reliable test of young children's or even infants' numerical abilities is developed using ERPs or NIRS, we may begin imaging populations of children at risk of developing dyscalculia, such as FAS or premature children.

Sequences of Calculation and Frontal Cortex

Once we begin to better understand the cerebral networks underlying simple calculations, and how they vary with age, education, and expertise, we can begin to address the issue of their coordination. Multidigit calculations such as 23 + 48 call for the sequential execution of several operations that are partially contingent on previous results (3 + 8 = 11, then 2 + 4 + carry 1 = 7). Prefrontal cortex and anterior cingulate cortex have been associated with the working memory and executive control systems that are needed for attentive, sequential task execution and task control. Indeed, these areas are active during serial subtraction problems (Chochon et al., 1999; Moonen et al., 1990; Roland and Friberg, 1985). I plan to study their specific contributions to calculation using a combination of brain-imaging and computational modeling.

In one experiment, subjects would perform different sequences of internal operations on the same pairs of sequentially presented digits: (a) subtracting the two digits (2 7 → "five"), thus engaging the quantity system; (b) multiplying the two digits (2 7 → "fourteen"), thus engaging rote memory for arithmetic facts; (c) subtracting, then multiplying (e.g., for the pair 2 7, first compute 7 − 2, then 2 × 7); (d) multiplying, then subtracting;

(e) subtracting, then multiplying *the obtained result* (e.g., $7 - 2 = 5$, then $5 \times 7 = 35$); (f) multiplying, then subtracting *the obtained result* (e.g., $2 \times 7 = 14$, then $14 - 7 = 7$). Activations would be imaged both with fMRI and with ERPs. The triple-code model predicts that ERPs should show the successive activation of parietal and left perisylvian networks in the task-imposed order. Furthermore, the sequence requirement (when having to execute two operations) and the working memory requirement (when having to hold an intermediate result online for further calculations) should yield a detectable additional activation in prefrontal and cingulate areas. Ultimately, other experiments could be designed to explore the contribution of anterior areas to more-complex operations such as those engaged when computing, say, 24×17 mentally.

Computational modeling of neural networks will also be used to explore the role of various prefrontal, cingulate, and basal subcortical circuits in executive functions. For the past 10 years, Changeux and I have developed neuronal models of functions associated with the prefrontal cortex. We have successively modeled the delayed-response task (Dehaene and Changeux, 1989), the Wisconsin Card Sorting Test (Dehaene and Changeux, 1991), the putative role of working memory in early numerical development (Dehaene and Changeux, 1993), and most recently the Tower of London Test (Dehaene and Changeux, 1997). The latter model provides a general architecture for the internal generation of a hierarchical sequence of actions aimed at achieving a certain goal. In the future, we hope to extend it to the case of simple arithmetic problems. The main advantage of this form of modeling is that it obliges us to specify in all detail the putative organization and connectivity patterns of functional subsystems such as planning, working memory, error correction, backtracking, or evaluation that are usually given only a loose definition in most "boxological" cognitive models. Hence, the model should provide more-detailed predictions for neuropsychological and brain-imaging studies.

Conscious and Unconscious Number Processing

Studies of number processing can also be relevant to more-general issues in cognitive neuroscience. A considerable controversy in human cognition concerns the extent to which a masked word may contact its semantic representation, or whether consciousness is necessary for semantic processing. Recently, with Ph.D. students Koechlin and Naccache, I have designed an experimental situation in which subjects demonstrably process up to a very high level a numerical information that they claim not to have seen (Dehaene et al., 1998c). Using this masked priming situation, I hope to bring new light to bear on the conditions under which information may or may not enter consciousness and to identify the relevant networks of cerebral areas using brain-imaging techniques.

Our experiment consists in presenting on a computer screen a random string of consonants (e.g., QfFmPg), followed by a number in Arabic or in verbal notation (the prime; e.g., "two" or "2"), then another consonant string, and finally a second number (the target; e.g., "four" or "4"). When the prime is presented for a very short duration such as 43 ms, subjects cannot report it and claim to be aware only of the second target number. When, however, they are asked to decide if the target is larger or smaller than 5, their response time varies with the prime. On congruent trials where both prime and target fall on the same side of 5 (both larger or both smaller), responses are faster than on incongruent trials when they fall on opposite sides of 5. The effect, which is totally independent of prime and target notation, proves that the prime number has been unconsciously identified, associated with a certain quantity, and categorized as larger or smaller than 5. Unlike most semantic priming experiments, the effect is extremely robust (20 ms; all subjects show a positive priming effect). It can be considered as a normal subject form of "blindsight" (Weiskrantz, 1997).

This experiment opens up remarkable possibilities for studying the cerebral bases of unconscious semantic processing. First, we can record ERPs during the task. Our results to date indicate that ERPs are sufficiently sensitive to detect changes in cerebral activity associated with the prime only. For instance, we detect distinct activation patterns for primes in Arabic or in verbal notation. We also detect differences between congruent and incongruent trials. We even detect a motor readiness potential contralateral to the hand that the subject should use if he had to respond to the prime. It thus becomes possible to follow the sequence of cerebral processing of the prime from visual to semantic to motor areas. We have shown that single-event fMRI can detect small but significant differences in activation as a function of whether the prime "votes" for the left or right hand, a result which confirms the ERP lateralized readiness potential effect (Dehaene et al., 1998c).

If brain imaging continues to prove sensitive to subliminal cerebral processes, this will open up new perspectives. We plan to replicate our experiments while varying the task. This should enable us to explore which tasks can or cannot be executed unconsciously on masked primes. We might thus define a taxonomy of tasks that do or do not require conscious guidance. Brain imaging might then tell us whether this dissociation can be related to the activation of specific anatomical areas such as the anterior executive system (prefrontal cortex, anterior cingulate). My recently developed "global workspace" model of the neuronal networks underlying conscious effortful tasks will provide a theoretical framework for this research (Dehaene et al., 1998a).

Numbers Among Other Categories of Words

One final issue I would like to address in future research is the notion of category-specific representations. As I mentioned in the introduction, numbers provide probably one of the best demonstrations of a biologically determined domain of knowledge, with evidence of precursor knowledge in animals and human infants, category-specific deficits in brain-lesioned patients, and a reproducible neural substrate as seen in brain imaging. In the next 10 years, I plan to extend the research strategy used with numbers to other categories of words. My colleagues shall study the core knowledge of several plausible categories of words, including animals, foods, colors, and numbers, but we shall focus particularly on the category of body parts, for which there is already some neuropsychological evidence for category-specific deficits (e.g., Dennis, 1976; Sirigu et al., 1991).

Behaviorally, we plan to run chronometric experiments with normal subjects aiming at determining the internal organization of the adult representation of the body. Suppose that subjects are asked to determine which of two body parts (say, knee and ankle) is higher—a task analogous to number comparison. We again expect to find a distance effect. Subjects should be faster for more-distant pairs, indicating that they have an internal representation of the body schema. Indeed, a distance effect on the body schema has been reported in a slightly different task setting (Lakatos and Shepard, 1997). By determining the precise characteristics of this distance metric, examining whether there is an automatic intrusion of a distance effect even in same–different judgments, and examining whether it depends on the input modality (name versus picture), we should be able to characterize this representation, its abstractness, and the automaticity of its access, just like we did with number.

Neuropsychologically, Cohen and I will be looking for cases of autotopoagnosia, or selective deficits of body knowledge, as well as cases of selective naming or comprehension deficits for body part names. As in our previous case studies in the number domain, we shall then characterize the processing level at which the deficit occurs by studying whether it generalizes to different input–output modalities and different semantic or nonsemantic tasks.

Finally, using brain imaging in adults, we shall also examine the presence of areas specifically activated during body part comparison as opposed to number comparison. In an experiment we are currently undertaking, 19 names of body parts are matched in frequency, length, and number of syllables to 19 number names. In different blocks, the words are presented auditorily or visually, and in English or in French, to bilingual subjects, while brain activity is recorded with 3-Tesla fMRI. Using a conjunction analysis in the statistical package SPM96, the results allow us to test the prediction that there are brain regions associated with number meaning or with body part meaning that are active only for a certain category of word, but are irrespective of the modality or language used to access that category. Our preliminary results do indeed show such activity in the intraparietal region for numbers and in the left parietal cortex for body parts. In the future, we shall also replicate these results with ERPs. By examining at what point the ERPs to different categories of words begin to diverge, we shall estimate how quickly the semantic representation is accessed in different modalities and languages of input. I have successfully used this approach in a previous ERP experiment in which I showed striking ERP differences to proper names, animal names, action verbs, and numerals by 250–300 ms following visual word presentation (Dehaene, 1995).

CONCLUSIONS AND IMPLICATIONS

The present project can have a significant impact on our understanding of brain function and on its application to pressing societal issues. At the fundamental level, it will provide an in-depth understanding of how the networks for number processing develop in infants and in young children and how they are organized in adults as a function of task and expertise. In the past 10 years, number processing has become a central domain of human cognition, with its specialized scientific journal (*Mathematical Cognition*, edited by Brian Butterworth) and several dedicated congresses. The triple-code model of number processing currently provides a useful integrative framework for this domain. I intend to continue developing my research on elementary arithmetic to make it one of the best demonstrations of a biologically determined, category-specific domain, with evidence ranging from infant and animal studies to neuropsychology, brain imaging, and neuronal modeling. Starting from the numerical field, forays will also be made into other central issues in cognitive science, including conscious and unconscious processing and the nature and origins of category-specific representations.

Calculation and arithmetic are also a domain of central importance to society. Dyscalculia is a frequent and poorly understood deficit in school children, with causes that range from neurological damage to an acquired phobia for mathematics. Even in children or adults who have followed a normal school curriculum, innumeracy is a growing concern. Although my research is mostly oriented toward the fundamental goal of understanding brain function, there may eventually be important applications in the domain of developmental dyscalculia, innumeracy, and their early diagnosis and rehabilitation. The triple-code model of number processing is already being applied by developmental psychologists such as Robbie Case to better understand normal and impaired numerical development. Cognitive tests are also being incorporated in batteries used for the diagnosis and rehabilitation of acalculia, not only in adults, but also in younger brain-lesioned patients. In the future, it seems likely that brain-imaging tools such as the ones I am helping develop will be used to understand the normal development of the cerebral networks for arithmetic and its variability and occasional failure. Once reliable imaging paradigms are developed, especially in young children and infants, they will be readily applicable to at-risk populations such as premature or FAS babies. They will also help understand and facilitate the recovery of adult patients with acalculia stemming from focal brain damage.

Finally, is it unreasonable to hope that, by improving our understanding of how the brain acquires mathematics, one may also discover better ways of teaching it? Quite the contrary, I believe that this is an achievable goal. Educational methods with a sound scientific basis are perhaps the most important contribution that cognitive neuroscience has to make to society. Our current understanding of numerical cognition already sheds some light on issues such as how one should teach the multiplication table or whether and how electronic calculators should be used in classrooms (Dehaene, 1997). Although these conclusions are still speculative, many educational developments are to be expected as the field of numerical cognition matures in the next 10 years.

REFERENCES

Abdullaev, Y. G., and K. V. Melnichuk. 1996. Counting and arithmetic functions of neurons in the human parietal cortex. *NeuroImage* 3:S216.

Antell, S. E., and D. P. Keating. 1983. Perception of numerical invariance in neonates. *Child Development* 54:695–701.

Ashcraft, M. H. 1992. Cognitive arithmetic: A review of data and theory. *Cognition* 44:75–106.

Ashcraft, M. H., and E. H. Stazyk. 1981. Mental addition: A test of three verification models. *Memory and Cognition* 9:185–196.

Barkow, J. H., L. Cosmides, and J. Tooby, eds. 1992. *The Adapted Mind: Evolutionary Psychology and the Generation of Culture.* New York: Oxford University Press.

Benson, D. F., and N. Geschwind. 1970. Developmental Gerstmann syndrome. *Neurology* 20:293–298.

Benton, A. L. 1987. Mathematical disability and the Gerstmann syndrome. In G. Deloche and X. Seron, eds. *Mathematical Disabilities: A Cognitive Neuropsychological Perspective.* Hillsdale, N.J.: Lawrence Erlbaum Associates.

Bijeljac-Babic, R., J. Bertoncini, and J. Mehler. 1991. How do four-day-old infants categorize multisyllabic utterances. *Developmental Psychology* 29:711–721.

Boysen, S. T., and G. G. Berntson. 1996. Quantity-based interference and symbolic representations in chimpanzees (*Pan trogolodytes*). *Journal of Experimental Psychology: Animal Behavior Processes* 22:76–86.

Boysen, S. T., and E. J. Capaldi. 1993. *The Development of Numerical Competence: Animal and Human Models.* Hillsdale, N.J.: Erlbaum.

Brannon, E. M., and H. S. Terrace. 1998. Ordering of the numerosities 1 to 9 by monkeys. *Science* 282(5389):746–749.

Brannon, E. M., and H. S. Terrace. 2000. Representation of the numerosities 1–9 by rhesus macaques (*Macaca mulatta*). *Journal of Experimental Psychology: Animal Behavior Processes* 26:31–49.

Brysbaert, M. 1995. Arabic number reading: On the nature of the numerical scale and the origin of phonological recoding. *Journal of Experimental Psychology: General* 124:434–452.

Buckley, P. B., and C. B. Gillman. 1974. Comparison of digits and dot patterns. *Journal of Experimental Psychology* 103:1131–1136.

Burbaud, P., P. Degreze, P. Lafon, J.-M. Franconi, B. Bouligand, B. Bioulac, J.-M. Caille, and M. Allard. 1995. Lateralization of prefrontal activation during internal mental calculation: A functional magnetic resonance imaging study. *Journal of Neurophysiology* 74:2194–2200.

Campbell, J. I. D., and J. M. Clark. 1988. An encoding complex view of cognitive number processing: Comment on McCloskey, Sokol & Goodman (1986). *Journal of Experimental Psychology: General* 117:204–214.

Capaldi, E. J., and D. J. Miller. 1988. Counting in rats: Its functional significance and the independent cognitive processes that constitute it. *Journal of Experimental Psychology: Animal Behavior Processes* 14:3–17.

Caramazza, A. 1996. The brain's dictionary. *Nature* 380:485–486.

Caramazza, A., and M. McCloskey. 1987. Dissociations of calculation processes. In G. Deloche and X. Seron, eds. *Mathematical Disabilities: A Cognitive Neuropsychological Perspective.* Hillsdale, N.J.: Lawrence Erlbaum Associates.

Caramazza, A., and J. R. Shelton. 1998. Domain-specific knowledge systems in the brain: The animate-inanimate distinction. *Journal of Cognitive Neuroscience* 10:1–34.

Chochon, F., L. Cohen, P. F. van de Moortele, and S. Dehaene. 1999. Differential contributions of the left and right inferior parietal lobules to number processing. *Journal of Cognitive Neuroscience* 11:617–630.

Church, R. M., and W. H. Meck. 1984. The numerical attribute of stimuli. In H. L. Roitblat, T. G. Bever, and H. S. Terrace, eds. *Animal Cognition.* Hillsdale, N.J.: Erlbaum.

Cipolotti, L., and B. Butterworth. 1995. Toward a multiroute model of number processing: Impaired number transcoding with preserved calculation skills. *Journal of Experimental Psychology: General* 124:375–390.

Cipolotti, L., B. Butterworth, and G. Denes. 1991. A specific deficit for numbers in a case of dense acalculia. *Brain* 114:2619–2637.

Cipolotti, L., B. Butterworth, and E. K. Warrington. 1994. From one thousand nine hundred and forty-five to 1000945. *Neuropsychologia* 32:503–509.

Cipolotti, L., E. K. Warrington, and B. Butterworth. 1995. Selective impairment in manipulating arabic numerals. *Cortex* 31:73–86.

Cohen, L., and S. Dehaene. 1991. Neglect dyslexia for numbers? A case report. *Cognitive Neuropsychology* 8:39–58.

Cohen, L., and S. Dehaene. 1994. Amnesia for arithmetic facts: A single case study. *Brain and Language* 47:214–232.

Cohen, L., and S. Dehaene. 1995. Number processing in pure alexia: The effect of hemispheric asymmetries and task demands. *NeuroCase* 1:121–137.

Cohen, L., and S. Dehaene, 1996. Cerebral networks for number processing: Evidence from a case of posterior callosal lesion. *NeuroCase* 2:155–174.

Cohen, L., S. Dehaene, and P. Verstichel. 1994. Number words and number non-words: A case of deep dyslexia extending to arabic numerals. *Brain* 117:267–279.

Dagenbach, D., and M. McCloskey. 1992. The organization of arithmetic facts in memory: Evidence from a brain-damaged patient. *Brain and Cognition* 20:345–366.

Damasio, H., T. J. Grabowski, D. Tranel, R. D. Hichwa, and A. R. Damasio. 1996. A neural basis for lexical retrieval. *Nature* 380:499–505.

Davis, H., and R. Pérusse. 1988. Numerical competence in animals: Definitional issues current evidence and a new research agenda. *Behavioral and Brain Sciences* 11:561–615.

Dehaene, S. 1989. The psychophysics of numerical comparison: A re-examination of apparently incompatible data. *Perception and Psychophysics* 45:557–566.

Dehaene, S. 1992. Varieties of numerical abilities. *Cognition* 44:1–42.

Dehaene, S. 1995. Electrophysiological evidence for category-specific word processing in the normal human brain. *NeuroReport* 6:2153–2157.

Dehaene, S. 1996. The organization of brain activations in number comparison: Event-related potentials and the additive-factors methods. *Journal of Cognitive Neuroscience* 8:47–68.

Dehaene, S. 1997. *The Number Sense*. New York: Oxford University Press.

Dehaene, S., and R. Akhavein. 1995. Attention, automaticity and levels of representation in number processing. *Journal of Experimental Psychology: Learning, Memory, and Cognition* 21:314–326.

Dehaene, S., and J. P. Changeux. 1989. A simple model of prefrontal cortex function in delayed-response tasks. *Journal of Cognitive Neuroscience* 1:244–261.

Dehaene, S., and J. P. Changeux. 1991. The Wisconsin Card Sorting Test: Theoretical analysis and modelling in a neuronal network. *Cerebral Cortex* 1:62–79.

Dehaene, S., and J. P. Changeux. 1993. Development of elementary numerical abilities: A neuronal model. *Journal of Cognitive Neuroscience* 5:390–407.

Dehaene, S., and J. P. Changeux. 1997. A hierarchical neuronal network for planning behavior. *Proceedings of the National Academy of Sciences of the United States of America* 94:13,293–13,298.

Dehaene, S., and L. Cohen. 1991. Two mental calculation systems: A case study of severe acalculia with preserved approximation. *Neuropsychologia* 29:1045–1074.

Dehaene, S., and L. Cohen. 1994. Dissociable mechanisms of subitizing and counting: Neuropsychological evidence from simultanagnosic patients. *Journal of Experimental Psychology: Human Perception and Performance* 20:958–975.

Dehaene, S., and L. Cohen. 1995. Towards an anatomical and functional model of number processing. *Mathematical Cognition* 1:83–120.

Dehaene, S., and L. Cohen. 1997. Cerebral pathways for calculation: Double dissociation between rote verbal and quantitative knowledge of arithmetic. *Cortex* 33:219–250.

Dehaene, S., and J. Mehler. 1992. Cross-linguistic regularities in the frequency of number words. *Cognition* 43:1–29.

Dehaene, S., E. Dupoux, and J. Mehler. 1990. Is numerical comparison digital?: Analogical and symbolic effects in two-digit number comparison. *Journal of Experimental Psychology: Human Perception and Performance* 16:626–641.

Dehaene, S., S. Bossini, and P. Giraux. 1993. The mental representation of parity and numerical magnitude. *Journal of Experimental Psychology: General* 122:371–396.

Dehaene, S., M. I. Posner, and D. M. Tucker. 1994. Localization of a neural system for error detection and compensation. *Psychological Science* 5:303–305.

Dehaene, S., N. Tzourio, V. Frak, L. Raynaud, L., Cohen, J. Mehler, and B. Mazoyer. 1996. Cerebral activations during number multiplication and comparison: A PET study. *Neuropsychologia* 34:1097–1106.

Dehaene, S., M. Kerszberg, and J. P. Changeux. 1998a. A neuronal model of a global workspace in effortful cognitive tasks. *Proceedings of the National Academy of Sciences of the United States of America* 95:14,529–14,534.

Dehaene, S., G. Dehaene-Lambertz, and L. Cohen. 1998b. Abstract representations of numbers in the animal and human brain. *Trends in Neuroscience* 21:355–361.

Dehaene, S., L. Naccache, G. Le Clec'H, E. Koechlin, M. Mueller, G. Dehaene-Lambertz, P. F. van de Moortele, and D. Le Bihan. 1998c. Imaging unconscious semantic priming. *Nature* 395:597–600.

Dehaene, S., E. Spelke, R. Stanescu, P. Pinel, and S. Tsivkin. 1999. Sources of mathematical thinking: Behavioral and brain-imaging evidence. *Science* 284:970–974.

Dehaene-Lambertz, G., and S. Dehaene. 1994. Speed and cerebral correlates of syllable discrimination in infants. *Nature* 370:292–295.

Deloche, G., and X. Seron. 1982a. From one to 1: An analysis of a transcoding process by means of neuropsychological data. *Cognition* 12:119–149.

Deloche, G., and X. Seron. 1982b. From three to 3: A differential analysis of skills in transcoding quantities between patients with Broca's and Wernicke's aphasia. *Brain* 105:719–733.

Deloche, G., and X. Seron. 1984. Semantic errors reconsidered in the procedural light of stack concepts. *Brain and Language* 21:59–71.

Deloche, G., and X. Seron. 1987. Numerical transcoding: A general production model. In G. Deloche and X. Seron, eds. *Mathematical Disabilities: A Cognitive Neuropsychological Perspective*. Hillsdale N.J.: Lawrence Erlbaum Associates.

Dennis, M.. 1976. Dissociated naming and locating of body parts after left anterior temporal lobe resection: An experimental case study. *Brain and Language* 3:147–163.

Duncan, E. M., and C. E. McFarland. 1980. Isolating the effects of symbolic distance and semantic congruity in comparative judgments: An additive-factors analysis. *Memory and Cognition* 8:612–622.

Fias, W., M. Brysbaert, F. Geypens, and G. d'Ydewalle. 1996. The importance of magnitude information in numerical processing: Evidence from the SNARC effect. *Mathematical Cognition* 2:95–110.

Fodor, J. A. 1983. *The Modularity of Mind*. Cambridge, Mass.: MIT Press.

Gallistel, C. R., 1989. Animal cognition: The representation of space, time, and number. *Annual Review of Psychology* 40:155–189.

Gallistel, C. R.. 1990. *The Organization of Learning*. Cambridge, Mass.: MIT Press.

Gallistel, C. R., and R. Gelman, 1992. Preverbal and verbal counting and computation. *Cognition* 44:43–74.

Galton, F. 1880. Visualised numerals. *Nature* 21:252–256.

Gazzaniga, M. S., and S. A. Hillyard. 1971. Language and speech capacity of the right hemisphere. *Neuropsychologia* 9:273–280.

Gazzaniga, M. S., and C. E. Smylie. 1984. Dissociation of language and cognition: A psychological profile of two disconnected right hemispheres. *Brain* 107:145–153.

Gerstmann, J.. 1940. Syndrome of finger agnosia disorientation for right and left agraphia and acalculia. *Archives of Neurology and Psychiatry* 44:398–408.

Grafman, J., D. Kampen, J. Rosenberg, A. Salazar, and F. Boller. 1989. Calculation abilities in a patient with a virtual left hemispherectomy. *Behavioural Neurology* 2:183–194.

Gratton, G., M. Fabiani, D. Friedman, M. A. Franceschini, S. Fantini, P. Corballis, and E. Gratton. 1995. Rapid changes of optical parameters in the human brain during a tapping task. *Journal of Cognitive Neuroscience* 7:446–456.

Hauser, M. D., P. MacNeilage, and M. Ware. 1996. Numerical representations in primates. *Proceedings of the National Academy of Sciences of the United States of America* 93:1,514–1,517.

Hécaen, H., R. Angelergues, and S. Houillier. 1961. Les variétés cliniques des acalculies au cours des lésions rétro-rolandiques: Approche statistique du problème. *Revue Neurologique* 105:85–103.

Henik, A., and J. Tzelgov. 1982. Is three greater than five?: The relation between physical and semantic size in comparison tasks. *Memory and Cognition* 10:389–395.

Henschen, S. E. 1920. *Klinische und Anatomische Beitraege zur Pathologie des Gehirns.* Stockholm: Nordiska Bokhandeln.

Hillis, A. E., and A. Caramazza. 1995. Representation of grammatical categories of words in the brain. *Journal of Cognitive Neuroscience* 7:396–407.

Hinrichs, J. V., D. S. Yurko, and J. M. Hu. 1981. Two-digit number comparison: Use of place information. *Journal of Experimental Psychology: Human Perception and Performance* 7:890–901.

Hoshi, Y., and M. Tamura. 1997. Near-infrared optical detection of sequential brain activation in the prefrontal cortex during mental tasks. *NeuroImage* 5:292–297.

Huha, E. M., D. B. Berch, and R. Krikorian. 1995. Obligatory Activation of Magnitude Information During Non-Numerical Judgments of Arabic Numerals. Poster presented at the annual meeting of the American Psychological Society, New York.

Inouye, T., K. Shinosaki, A. Iyama, and Y. Mastumoto. 1993. Localisation of activated areas and directional EEG patterns during mental arithmetic. *Electroencephalography and Clinical Neurophysiology* 86:224–230.

Kiefer, M., and S. Dehaene. 1997. The time course of parietal activation in single-digit multiplication: Evidence from event-related potentials. *Mathematical Cognition* 3:1–30.

Kleinschmidt, A., H. Obrig, M. Requardt, K. D. Merboldt, U. Dirnagl, A. Villringer, and J. Frahm. 1996. Simultaneous recording of cerebral blood oxygenation changes during human brain activation by magnetic resonance imaging and near-infrared spectroscopy. Journal of Cerebral Blood Flow and Metabolism 16:817–826.

Koechlin, E., S. Dehaene, and J. Mehler. 1997. Numerical transformations in five month old human infants. *Mathematical Cognition* 3:89–104.

Kopera-Frye, K., S. Dehaene, and A. P. Streissguth. 1996. Impairments of number processing induced by prenatal alcohol exposure. *Neuropsychologia* 34:1187–1196.

Lakatos, S., and R. N. Shepard. 1997. Time-distance relations in shifting attention between locations on one's body. *Perception and Psychophysics* 59(4):557–566.

LeFevre, J., J. Bisanz, and L. Mrkonjic. 1988. Cognitive arithmetic: Evidence for obligatory activation of arithmetic facts. *Memory and Cognition* 16:45–53.

Macaruso, P., M. McCloskey, and D. Aliminosa. 1993. The functional architecture of the cognitive numerical-processing system: Evidence from a patient with multiple impairments. *Cognitive Neuropsychology* 10:341–376.

Marr, D. 1982. *Vision: A Computational Investigation into the Human Representation and Processing of Visual Information.* New York: W. H. Freeman.

Martin, A., J. V. Haxby, F. M. Lalonde. C. L. Wiggs, and L. G. Ungerleider. 1995. Discrete cortical regions associated with knowledge of color and knowledge of action. *Science* 270:102–105.

Martin, A., C. L. Wiggs, L. G. Ungerleider, and J. V. Haxby. 1996. Neural correlates of category-specific knowledge. *Nature* 379:649–652.

Matsuzawa, T. 1985. Use of numbers by a chimpanzee. *Nature* 315:57–59.

McCarthy, R. A., and E. K. Warrington. 1988. Evidence for modality-specific meaning systems in the brain. *Nature* 334:428–430.

McCarthy, R. A., and E. K. Warrington. 1990. *Cognitive Neuropsychology: A Clinical Introduction.* San Diego, Calif.: Academic Press.

McCloskey, M. 1992. Cognitive mechanisms in numerical processing: Evidence from acquired dyscalculia. *Cognition* 44:107–157.

McCloskey, M. 1993. Theory and evidence in cognitive neuropsychology: A radical response to Robertson, Knight, Rafal, and Shimamura (1993). *Journal of Experimental Psychology: Learning* 19:718–734.

McCloskey, M., and A. Caramazza. 1987. Cognitive mechanisms in normal and impaired number processing. In G. Deloche and X. Seron, eds. *Mathematical Disabilities: A Cognitive Neuropsychological Perspective.* Hillsdale, N.J.: Lawrence Erlbaum Associates.

McCloskey, M., A. Caramazza, and A. Basili. 1985. Cognitive mechanisms in number processing and calculation: Evidence from dyscalculia. *Brain and Cognition* 4:171–196.

McCloskey, M., S. M. Sokol, and R. A. Goodman. 1986. Cognitive processes in verbal-number production: Inferences from the performance of brain-damaged subjects. *Journal of Experimental Psychology: General* 115:307–330.

McCloskey, M., D. Aliminosa, and S. M. Sokol. 1991a. Facts, rules and procedures in normal calculation: Evidence from multiple single-patient studies of impaired arithmetic fact retrieval. *Brain and Cognition* 17:154–203.

McCloskey, M., W. Harley, and S. M. Sokol. 1991b. Models of arithmetic fact retrieval: An evaluation in light of findings from normal and brain-damaged subjects. *Journal of Experimental Psychology: Learning* 17:377–397.

McCloskey, M., P. Macaruso, and T. Whetstone. 1992. The functional architecture of numerical processing mechanisms: Defending the modular model. In J. I. D. Campbell, ed. *The Nature and Origins of Mathematical Skills.* Amsterdam: Elsevier.

Mechner, F. 1958. Probability relations within response sequences under ratio reinforcement. *Journal of the Experimental Analysis of Behavior* 1:109–121.

Mechner, F., and L. Guevrekian. 1962. Effects of deprivation upon counting and timing in rats. *Journal of the Experimental Analysis of Behavior* 5:463–466.

Meck, W. H., and R. M. Church. 1983. A mode control model of counting and timing processes. *Journal of Experimental Psychology: Animal Behavior Processes* 9:320–334.

Meek, J. H., C. E. Elwell, M. J. Khan, J. Romaya, J. S. Wyatt, D.T. Delpy, and S. Zeki. 1995. Regional changes in cerebral haemodynamics as a result of a visual stimulus measured by near-infrared spectroscopy. *Proceedings of the Royal Society of London* Series B 261:351–356.

Mehler, J., and E. Dupoux. 1993. *What Infants Know. The New Cognitive Science of Early Development.* Oxford, U.K.: Blackwell.

Mitchell, R. W., P. Yao, P. T. Sherman, and M. O'Regan. 1985. Discriminative responding of a dolphin (*Tursiops truncatus*) to differentially rewarded stimuli. *Journal of Comparative Psychology* 99:218–225.

Moonen, C. T. W., P. C. M. Vanzijl, J. A. Frank, D. LeBihan, and E. D. Becker. 1990. Functional magnetic resonance imaging in medicine and physiology. *Science* 250:53–61.

Moore, D., J. Benenson, J. S. Reznick, M. Peterson, and J. Kagan. 1987. Effect of auditory numerical information on infant's looking behavior: Contradictory evidence. *Developmental Psychology* 23:665–670.

Moyer, R. S., and T. K. Landauer. 1967. Time required for judgements of numerical inequality. *Nature* 215:1519–1520.

Näätänen, R. 1990. The role of attention in auditory information processing as revealed by event-related potentials and other brain measures of cognitive function. *Behavioral and Brain Sciences* 13:201–288.

Noel, M. P., and X. Seron. 1993. Arabic number reading deficit: A single case study. *Cognitive Neuropsychology* 10:317–339.

Pepperberg, I. M. 1987. Evidence for conceptual quantitative abilities in the African grey parrot: Labeling of cardinal sets. *Ethology* 75:37–61.

Pinker, S. 1995. *The Language Instinct.* London: Penguin.

Pinker, S. 1997. *How the Mind Works.* New York: Norton.

Platt, J. R., and D. M. Johnson. 1971. Localization of position within a homogeneous behavior chain: Effects of error contingencies. *Learning and Motivation* 2:386–414.

Restle, F. 1970. Speed of adding and comparing numbers. *Journal of Experimental Psychology* 91:191–205.

Rilling, M., and C. McDiarmid. 1965. Signal detection in fixed ratio schedules. *Science* 148:526–527.

Roland, P. E., and L. Friberg. 1985. Localization of cortical areas activated by thinking. *Journal of Neurophysiology* 53:1219–1243.

Rueckert, L., N. Lange, A. Partiot, I. Appollonio, I. Litvar, D. Le Bihan, and J. Grafman. 1996. Visualizing cortical activation during mental calculation with functional MRI. *NeuroImage* 3:97–103.

Rumbaugh, D. M., S. Savage-Rumbaugh, and M. T. Hegel. 1987. Summation in the chimpanzee (*Pan troglodytes*). *Journal of Experimental Psychology: Animal Behavior Processes* 13:107–115.

Seron, X., and G. Deloche. 1983. From 4 to four: A supplement to from three to 3. *Brain* 106:735–744.

Seron, X., M. Pesenti, M. P. Noël, G. Deloche, and J.-A. Cornet. 1992. Images of numbers or when 98 is upper left and 6 sky blue. *Cognition* 44:159–196.

Seymour, S. E., P. A. Reuter-Lorenz, and M. S. Gazzaniga. 1994. The disconnection syndrome: Basic findings reaffirmed. *Brain* 117:105–115.

Simon, T. J., S. J. Hespos, and P. Rochat. 1995. Do infants understand simple arithmetic? A replication of Wynn (1992). *Cognitive Development* 10:253–269.

Sirigu, A., J. Grafman, K. Bressler, and T. Sunderland. 1991. Multiple representations contribute to body knowledge processing: Evidence from a case of autotopoagnosia. *Brain* 114:629–642.

Sokol, S. M., R. Goodman-Schulman, and M. McCloskey. 1989. In defense of a modular architecture for the number-processing system: Reply to Campbell and Clark. *Journal of Experimental Psychology: General* 118:105–110.

Sokol, S. M., M. McCloskey, N. J. Cohen, and D. Aliminosa. 1991. Cognitive representations and processes in arithmetic: Inferences from the performance of brain-damaged subjects. *Journal of Experimental Psychology: Learning* 17:355–376.

Sophian, C., and N. Adams. 1987. Infants' understanding of numerical transformations. *British Journal of Developmental Psychology* 5:257–264.

Spellacy, F., and B. Peter. 1978. Dyscalculia and elements of the developmental Gerstmann syndrome in school children. *Cortex* 14:197–206.

Starkey, P., and R. G. Cooper. 1980. Perception of numbers by human infants. *Science* 210:1033–1035.

Starkey, P., E. S. Spelke, and R. Gelman. 1983. Detection of intermodal numerical correspondences by human infants. *Science* 222:179–181.

Starkey, P., E. S. Spelke, and R. Gelman. 1990. Numerical abstraction by human infants. *Cognition* 36:97–127.

Strauss, M. S., and L. E. Curtis. 1981. Infant perception of numerosity. *Child Development* 52:1146–1152.

Takayama, Y., M. Sugishita, I. Akiguchi, and J. Kimura. 1994. Isolated acalculia due to left parietal lesion. *Archives of Neurology* 51:286–291.

Temple, C. M. 1989. Digit dyslexia: A category-specific disorder in development dyscalculia. *Cognitive Neuropsychology* 6:93–116.

Temple, C. M. 1991. Procedural dyscalculia and number fact dyscalculia: Double dissociation in developmental dyscalculia. *Cognitive Neuropsychology* 8:155–176.

Thompson, R. F., K. S. Mayers, R. T. Robertson, and C. J. Patterson. 1970. Number coding in association cortex of the cat. *Science* 168:271–273.

Tzelgov, J., J. Meyer, and A. Henik. 1992. Automatic and intentional processing of numerical information. *Journal of Experimental Psychology: Learning, Memory, and Cognition* 18:166–179.

Ungerleider, L. G., and M. Mishkin. 1982. Two cortical visual systems. In D. J. Ingle, M. A. Goodale, and R. J. Mansfield, eds. *Analysis of Visual Behavior.* Cambridge, Mass.: MIT Press.

van Loosbroek, E., and A. W. Smitsman. 1990. Visual perception of numerosity in infancy. *Developmental Psychology* 26:916–922.

van Oeffelen, M. P., and P. G. Vos. 1982. A probabilistic model for the discrimination of visual number. *Perception and Psychophysics* 32:163–170.

Warrington, E. K. 1982. The fractionation of arithmetical skills: A single case study. *Quarterly Journal of Experimental Psychology* 34A:31–51.

Washburn, D. A., and D. M. Rumbaugh. 1991. Ordinal judgments of numerical symbols by macaques (*Macaca mulatta*). *Psychological Science* 2:190–193.

Weiskrantz, L. 1997. *Consciousness Lost and Found: A neurophyschological exploration.* New York: Oxford University Press.

Woodruff, G., and D. Premack. 1981. Primative (sic) mathematical concepts in the chimpanzee: Proportionality and numerosity. *Nature* 293:568–570.

Wynn, K. 1992a. Addition and subtraction by human infants. *Nature* 358:749–750.

Wynn, K. 1992b. Children's acquisition of the number words and the counting system. *Cognitive Psychology* 24:220–251.

Wynn, K. 1995. Origins of numerical knowledge. *Mathematical Cognition* 1:35–60.

Wynn, K. 1996. Infants' individuation and enumeration of actions. *Psychological Science* 7:164–169.

Xu, F., and S. Carey. 1996. Infants' metaphysics: The case of numerical identity. *Cognitive Psychology* 30:111–153.

Xu, F., and E. S. Spelke. 2000. Large number of discrimination in 6-month-old infants. *Cognition* 74(1):B1–B11.

Zbrodoff, N. J., and G. D. Logan. 1986. On the autonomy of mental processes: A case study of arithmetic. *Journal of Experimental Psychology: General* 115:118–130.

3

More Than Mere Coloring: A Dialog Between Philosophy and Neuroscience on the Nature of Spectral Vision

Kathleen A. Akins

INTRODUCTION

In its broadest form, the aim of my research is to bring together two disciplines with a shared intellectual territory—to find out what the combined resources of the neurosciences and philosophy can tell us about the nature of mind and its relation to the world.

In one direction, in bringing neuroscience to philosophy, my goal is to understand what empirical insights the new science can bring to a variety of traditional (and traditionally intransigent) philosophical problems, to the problems of our sensory attachment to the external world, of the self and its relation to the body, of conscious experience, and of mental representation in general. Like most philosophers, I assume that insofar as questions about the mind are empirical questions, the neurosciences will eventually provide empirical answers. This after all is the very purpose of the neurosciences, and as such is not surprising. A far more interesting possibility, however, is that neuroscience may well change the very questions asked, something that could occur in at least two different ways. First, a lesson I take from Quine (1960) is that as philosophers we are unlikely, in advance of the neurosciences, to have divided the problem space correctly, into questions that are amenable to empirical solution and those that are not—"these are the conceptual questions, impervious to the advances of science" and "these are the empirical questions, ripe for sci-

Department of Philosophy, Simon Fraser University, Burnby, British Columbia

entific explanation/revision." We should expect that as neuroscience advances it will jog the boundaries between the conceptual and the empirical as currently perceived. Second, following Sellars (1956), I suspect that our own mental events are, in some crucial respects, not unlike the postulated "unobservables" of scientific theories; at the very least, we do not have unmediated, "transparent" introspective access to our own psychological processes. But if this is so, perhaps in certain crucial respects the categories or classifications assumed by our everyday attributions of psychological events may not capture the mind's natural ontology. Perhaps the neurosciences will posit categories of neural events that cross cut the categories of common-sense psychology, such that the old questions no longer stand. Change the ontology of the mental, and you may change the shape of the problem space.

What then of the other direction, of the relation of philosophy to neuroscience? First, although the philosophy of neuroscience is as yet mostly uncharted territory, philosophy has a clear contribution to make in understanding the methodology of this essentially new science.

There is, however, another much more ambitious contribution that philosophy can make, one alluded to above in speaking of the "shared intellectual territory" of philosophy and neuroscience, namely, the interpretation of neuroscientific results. That is, I take quite seriously the view that the neurosciences and the philosophy of mind do have a common project, that the traditional philosophical problems about the mind and its relation to the world, are, at bottom, the same questions that animate both disciplines—and hence, that a neuroscientist is no more likely than a philosopher to know, for example, what a mental representation is, for this is (just) one of the common, central questions at issue. Indeed, in light of the reams of neuroscientific data already amassed (think here, for example, of the gigabytes of data from single-cell recordings that electrophysiologists have accumulated), one or two good ideas would go a long way toward meaningful interpretation of that data—and I suspect that several thousand years of philosophical thought about the nature of concepts, mental representation, sensory processing, and the self might well provide a few such insights.

That, in the abstract, is how I think things ought to work. The (million dollar) question is how to establish, concretely, the theoretical connections outlined above. Unlike many researchers, I do not expect that the problems of the mind/brain will be solved in one large leap, tamed, as it were, by The Great Idea. Indeed, I doubt that any of the central questions about the nature of mental representation, the self, or consciousness will each admit of a single solution. In light of this, I have adopted a piecemeal approach in my research: I have set myself a series of (what I hope will be) more tractable problems, each one focusing on a different area of visual neuroscience, which I hope will help me to untangle the mysteries of the mind/brain.

Below I illustrate this project by choosing one small puzzle, color vision—or what seems like a small problem only until one considers its long philosophical history and the recent explosion of color research in the neurosciences. By appealing to only widely accepted neuroscientific results, I hope to show how one might reinterpret the recent color research in a way that differs from the widely accepted version—and do so in a way that causes our understanding of our experience and phenomenology of color to change.

SPECTRAL VISION

A Puzzle Case

Let me begin by posing a puzzle about color vision and our conscious experience of color—or rather the case of M.S., an achromatopsic patient described by Heywood et al. (1994). To have cerebral achromatopsia is to suffer loss of color vision due to brain injury, a deficit that has been associated with damage to ventral occipital cortex (Damasio et al., 1980; Meadows, 1974). Typically, achromatopsic patients report that "their world looks dull, dirty, faded, gray, washed out, or devoid of color like a black and white TV" (Rizzo et al., 1992), and the standard tests for color perception bear this out—they test at chance or little better. When asked to name color samples, they respond with inappropriate color terms (e.g., in Rizzo et al., 1993, one subject identified a yellow sample as "pink" "because it is a lightish color"); when asked to name the typical colors of certain objects, they respond incorrectly (e.g., the same subject identified a banana as "green" because "it's a plant (and) most plants are green"). In the Heywood et al. (1994) study, the subject M.S. had bilateral ventral and ventromedial damage to the temporo-occipital regions and was classified as "densely achromatopsic:" He scored at chance on the Farnsworth-Munsell 100-Hue Test (which tests the just noticeable difference between 85 hues in a color circle); he failed to identify any of the concealed figures in the Ishihara pseudo-isochromatic plates (Figure 1); he was unable to discriminate between any two colors matched for brightness; and, like achromatopsics in general, he used color terms incorrectly and reported that the world appeared "gray."

In the reported experiments, however, M.S. showed certain specific responses to color stimuli. For example, given the stimulus in Figure 2a, a checkerboard of squares with low-luminosity contrast and a randomly placed light red square, M.S. could not pick out the red square. Yet, in Figure 2b, when a highly saturated red square was used, M.S. could pick

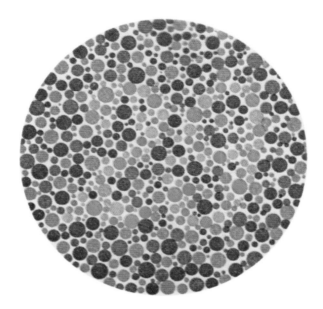

FIGURE 1   An Ishihara pseudo-isochromatic plate.

out "the different one." In Figure 2c, however, when the luminosity contrast of the background squares was increased, M.S. could no longer detect the saturated square.

At first glance, the explanation for this failure would seem to be the following. What allows M.S. to select the "odd man out" in Figure 2b is that he can see—in addition to luminance contrast—chromatic contrast when the colors are "deep" or saturated. What he cannot distinguish between are the two kinds of contrast, one from the other. Rather, what is signaled is merely "contrast of some type." Thus in Figure 2c, when there is both high luminance contrast between all of the squares and a high chromatic contrast between the red square and its neighbors, there is nothing to distinguish the red square. Every square contrasts with its neighbors in one way or another.

What makes this case puzzling, however, is that in further experiments the patient was able to detect and identify a colored form against a checkerboard background, whether that form was composed of saturated or unsaturated colored squares and whether the checkerboard background was of high or low luminance contrast. For example, the patient could not detect, in Figure 2c, the small, red square but he could detect

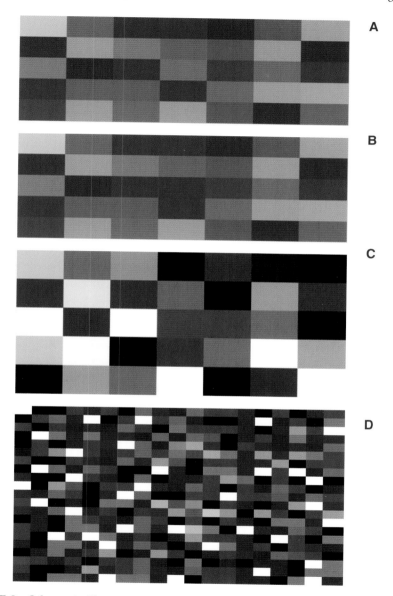

FIGURE 2   Schematic illustration of the stimuli used in Heywood et al. (1994).
A. Low luminance contrast checkerboard with desaturated red square. B. Low
luminance contrast checkerboard with highly saturated red square. C. High luminance contrast checkerboard with highly saturated red square. D. High luminance contrast checkerboard with desaturated red square and desaturated green
cross.

and trace with a finger the green cross in Figure 2d. He could also detect
and identify the red forms in Figure 2d. (This is particularly strange when
one recalls that M.S. was not able to detect or identify the concealed figures in the Ishihara plates.) Indeed, M.S. was able to pick out the individual squares that made up each of the forms—squares of exactly the
same size and saturation as those that, in isolation, he could not see. To
add to the mystery, further experiments showed that M.S. was also able to
see (apparent) motion given only chromatic information; and in brightness additivity tests, M.S. responded as would the normal subject, which
suggests that, at some level, color opponent processing remained intact.

The above results pose a number of difficult puzzles. For physiological theories of color vision, one puzzle—the one posed by Heywood et al.
at the end of their article—seems to be this. With bilateral damage to the
cortical "color center" (Heywood et al., 1994), M.S. has no phenomenal
experience of color, but he retains some color abilities but not others. What
residual chromatic information is available, then, and where is it available
in M.S.'s visual system? Why is it that exactly the same color information
is available for some tasks but not others (e.g., why can a colored square
be seen when it forms part of larger form but not when it is shown by
itself?).

For philosophical theories of color, the puzzle is not about M.S.'s chromatic information per se but about his phenomenological experience:
What is it like to be M.S. as he stares at the test stimuli? Here the problem
is not merely difficult but it is peculiar. In the color discrimination experiments that used single colored squares against the checkerboard background, it seems plausible that M.S. sees saturated chromatic contrast as
luminous contrast or as variations along a "gray scale." If, say, M.S. sees
the saturated red square as a very dark gray or black, then he would be
able to pick out the red square against a checkerboard background of low
luminosity contrast (i.e., he sees a black square amid a checkerboard of
varying shades of gray). This would also explain why, under the condition of high luminosity contrast, the red square became invisible: It would
be just one black square among many. Here again, the experiments using
colored forms show that this hypothesis could not be right, at least in
general. M.S. is able to pick out (flawlessly) colored forms against a high
luminosity contrast checkerboard, so, in some sense, he must be able to
see both luminance and chromatic contrast as well as the difference between them. Yet M.S. does not see the colored form as a *colored* form. In
what, then, does the phenomenal difference between luminance and chromaticity consist?

More to the point, what *could* the difference be given a world that
appears in shades of gray, a phenomenal space of one dimension? The
problem here is that the common-sense options for describing M.S.'s phe-

nomenology seem entirely inadequate. First, intuitively, we seem faced with only two possibilities:

1. Either M.S. sees in black and white or he sees in color.

We can understand, of course, that a person might have a limited color capacity, that he or she might be unable to discriminate between or categorize all of the hues perceived by the normal (human) observer. The red-green dichromate is an obvious and common example: For M.S., a pale green teacup and a pink one are indistinguishable by means of color. What seems nonsensical, however, is that a person might be able to see the world's objects "in color" yet at the same time perceive them as devoid of color or "in black and white." Second, in this case:

2. M.S. sees in black and white.

Or, at least, so it seems. Certainly this is what M.S. claims, and there is no reason to believe that he is insincere or in a state of general confusion. It is still more difficult to understand just how he might be genuinely mistaken about his own experiences. So, prima facie, M.S. sees in black and white. Hence:

3. M.S. does not see in color.

But M.S. must be able to see in color: if he could see only shades of gray, there would be no visual quality that distinguished the colored forms from the background. Thus, the paradox.

Recasting the Problem of Color Vision

When I open my eyes and look at my surroundings, I see the kitchen chair before me painted a uniform "colonial blue," the worn terra-cotta floor tiles in shades of rust; and, out the window, a dark green fig tree, now lit a brilliant yellow-green by the setting sun. On the common view, that is, to see "in color," to have human color vision, is just to see various media (the surface of objects, as well as gases, liquids, etc.) as colored; and this, in turn, means that we see each medium as having a particular determinate color or shade, a color which is of some type or category, and a color that is more or less independent of my present perception of it.

Looking at the chair, for example, I see "that exact color," a particular shade of blue that differs from the blue of the teapot on the table. In this sense of "color," "the exact color of the kitchen chair," it is estimated that we can see over a million colors or, more precisely, that we can make over a million discriminations within the range of chromaticity and lightness that define our phenomenal color space (Boynton, 1990, p. 238). This color phenomenology is said to be organized along three coordinates—hue, saturation, and lightness—or, alternatively, along the three "opponent" dimensions of lightness/darkness, red/green and blue/yellow. Intuitively, some of these colors appear to be "pure" or unitary colors (e.g., a pure red contains no yellow or blue), whereas others are "binary" hues apparently composed of two different hues (e.g., orange seems to be composed of red and yellow, whereas lime green is predominantly "pure" green with some yellow added).

In addition to seeing some particular color or other, we also categorize the colors. I do not see the chair as merely some determinate color or other, as "that one, there," but as a particular shade of blue. It is now widely accepted that there are basic color categories that are universally applied regardless of language or culture, although there is some disagreement about just which categories these are. As early as 1878, Hering suggested that there were but four elemental hue sensations—red, green, yellow, and blue—results confirmed by Sternheim and Boynton (1966) and Fuld et al. (1981). By contrast, Berlin and Kay (1969) concluded that there were 11 basic color terms common to (and translatable between) nearly 100 languages—in English these are red, yellow, green, blue, orange, purple, brown, pink, white, gray, and black, and these findings were confirmed in psychophysical experiments by Boynton and Olson (1987). Whatever the actual categories, Figure 3 shows one attempt, the Inter-Society Color Council–National Bureau of Standards (ISCC–NBS) method, to divide our continuous phenomenal space of color into the color categories.

Third, color is seen as an "inherent" property of objects (and other media), a property that objects have more or less independently of our present perceptions of them or, to use the philosophical terminology, for which an "appearance/reality" distinction holds. For example, looking out at the fig tree in the setting sun, I distinguish the leaves' present appearance, a luminescent lime, from their true color, matte dark green. I see the contribution of the light source—the warm golden light shining through and reflected from the leaves—as distinct from the color of the leaves themselves. Were I to watch the same scene through yellow-tinted sunglasses, I would distinguish between the true color of the leaves and their yellow appearance. Thus, naively at least, we distinguish between the colors that objects possess, in and of themselves, and the colors that objects often merely appear to have in virtue of external factors such as lighting or of our own subjective states. We see and conceive of colors as they are in themselves.

For the most part, research on human color vision—be that investigation philosophical, psychological, or physiological—has attempted to understand and explain this color phenomenology, our experience of a colored world. It is standard for philosophers and psychologists, for example, to ask about the evolution of human color vision and the historical function of seeing "in color." What survival advantage did seeing objects (and other media) as colored confer about our species—why do we see

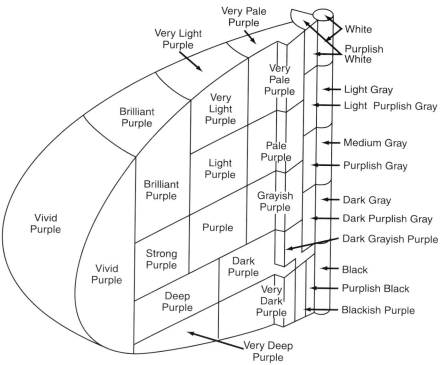

FIGURE 3 Illustration of a section of the ISCC–NBS color name chart in the purple region. The system divides the space into 267 blocks. The system uses the 10 different hue categories of Berlin and Kay, plus an additional one, "olive." These hue categories are divided through the use of modifiers—"pale," "light," "strong," "dark," "deep," "vivid," "brilliant" as well as the adverb "very" (from Kelly and Judd, 1976).

"in color"? Why do we have the particular color categories that we do— the reds, greens, yellows, and blues? Do our color categories serve to track salient features of our environment? And why does our phenomenal color space have the properties it does: Why does it have three dimensions, why are there "unitary" and "binary" colors, and so on (e.g., Shepard, 1997)? Computer scientists and psychophysicists, on the other hand, ask about which visual processes use color and for what purposes. Does seeing in color help us to locate objects of interest within a scene (e.g., to see the green pen amidst a jumble of office supplies in the desk drawer)? Does noting the color of an object help us to reidentify particulars (e.g, to locate your car among all the others in the parking lot) (e.g., Swain and Ballard, 1991)? Finally, neurophysiologists have asked about the physical realization of our color experience (e.g., Zeki, 1980). If the purpose of color vision is to produce "the colors," spectral processing must eventuate in a place where "the colors" are represented and which thereafter provides information about the colors (so encoded) to other visual areas. Thus we can expect to find in visual cortex a neural realization of our phenomenal color space, one that represents each determinate color as a single, determinate location within some color coordinate system, one that will explain the various phenomenal features of our particular color experience (e.g., its apparent three dimensionality and the unitary and binary hues). Exactly how such an area will be organized remains an open question: One need not expect to find the ISCC–NBS color system ("very dark olive") instantiated at a particular place in the gray matter. But given the phenomenology of color, some coordinate system, which maps onto our phenomenal color space, must exist—or so it has been commonly assumed.

Below, I make a rather odd suggestion about human color vision, namely, that the primary function of color vision is not to see the colors or to see various media as colored—to see, for example, ripe strawberries as red, the unclouded sky as blue, or the sun as bright orange. Rather, my suspicion is that color vision has evolved to encode a certain kind of general information about the retinal image, not color per se but *spectral contrast* information. Just as the luminance system encodes luminance contrast information for the purpose of discerning objects, properties, and events, the color system encodes spectral contrast information for much the same ends. Of course, we both perceive and conceive of various media as being colored, so spectral information must be used in this way as well. But seeing the colors is not the *central* function of color vision any more than seeing objects as bright or dark is the central purpose of luminance vision.

To explicate this view, I begin by setting aside the standard research questions above, questions that are driven by our first-person experience of a colored world. What would happen if we asked not about *color* per se as presented in our ordinary phenomenology—about the evolutionary history, function, and neural realization of color? What if we asked, instead, "What uses might a visual system make of *spectral information*?" If you think about "the world of light," that is, it is obvious that light, viewed as a waveform, has two dimensions, wavelength and amplitude. Normally, as researchers in vision, we concentrate on luminance information, on the amount of light independently of wavelength—on vision in a "black and white" world, as it were. Of course, our world is not a black and white world and the "spectral world," as such, is an exceedingly complex one in its own right.

Sunlight, as it filters through the atmosphere of earth, undergoes ozone absorption, molecular and aerosol scattering, plus water and oxygen absorption. When it reaches the ground, sunlight ranges in wavelength from about 300 nm on out past 1500 nm, with its peak intensity between 500 and 600 nm (Hendersen, 1977). Not coincidentally, as Boynton (1990) points out, the range of visible light for humans coincides with the range of greatest sunlight intensity, from 320 nm to around 800 nm, and its spectral sensitivity mimics the intensity curve of terrestrial sunlight, peaking at about 550 nm. Moreover, although we think of sunlight as constant in spectral composition (even though there may be more or less of it), in fact the spectral power distribution (SPD) of sunlight changes markedly throughout the day. This is intuitively obvious when we think of the quality of light at dawn (the "cool" hues), at midday (glaring "white"), and at dusk (the warm red-yellow hues). The SPD of sunlight is also dramatically affected by one's position relative to the Sun. Figure 4 shows, for example, the difference in SPDs reflected from an object facing toward and away from the Sun. All things being equal, a yellow banana facing away from the Sun ought to appear green, given the shift in SPD toward the short or blue end of the spectrum (Walraven et al., 1990).

The real complexity of the "spectral world," however, begins with the interaction of light with various media—the fact that light sources, in all their initial complexity, are absorbed, refracted, and reflected by a vast range of media and surfaces—and, as a result, the spectral composition and directionality of the light are altered in an indefinite number of ways. Take for example the media of air and water. Sunlight shining through a clear sky is scattered by the small particles of the atmosphere, a process that effects the short "blue" wavelengths selectively (hence the blue of the sky) but that leaves the overall directionality of the sunlight largely unchanged. On cloudy days, all wavelengths are scattered by the suspended water particles, yielding a diffuse illumination from all parts of the sky, and thus a reduction in the shadows of objects. In contrast, when sunlight shines on water, white light glints from the surface: Initially, some light in all wavelengths is reflected, a phenomenon known as specular reflectance. Most of the sunlight, however, enters the water and is absorbed by it: The deeper the water, the greater its absorption. Which wavelengths are absorbed, however, depends on the type of suspended organic particles in the water. For example, clear water with little organic matter will most easily absorb red and violet light, leaving only the intermediate "blue" wavelengths, whereas in marshes and swamps, the little light that is not absorbed by decomposing plants, tannins, lignins, and yellow-green plankton is in the red-orange region of the spectrum—hence the shimmering blue of the Mediterranean and the dark red-brown appearance of a swamp. Of course, water molecules (and the suspended particles in

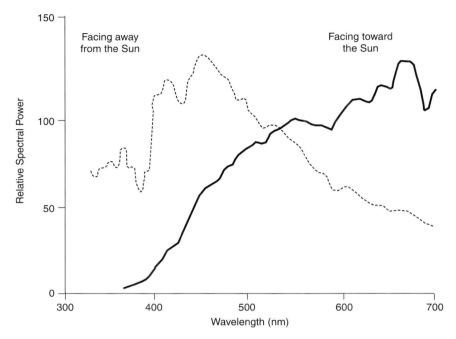

FIGURE 4   The relative SPDs of the light reflected from a single piece of white paper. The leftmost curve illustrates the SPD of the paper when illuminated by the "blue" light of morning, with the surface facing away from the Sun (at 30 degrees solar altitude); the rightmost curve illustrates the SPD of the same paper, illuminated with the "red" light of evening and with the surface facing toward the Sun (at 8 degrees solar altitude) (from Walraven et al., 1990)

water) also scatter light waves, so the medium of transport is itself "colored" (Levine and MacNichol, 1979; Munz and MacFarland, 1977).

Add to this the opaque and transparent objects of the world. Take, for example, the simplest case, pieces of uniformly dyed, matte construction paper (Figure 5). Looking at these spectral reflectance curves, it seems safe to say that our access to the spectral world, as human color observers, gives us very little insight into the spectral complexity of even a "uniform" surface color: Orange construction paper does not reflect one wavelength but rather its "spectral reflectance profile" is a continuous function across the range of visible light. Note that spectral reflectance curves give the *percentage* of light reflected at each wavelength (in the visible range), a measure that is constant across illuminations. But shine, say, a "red" light

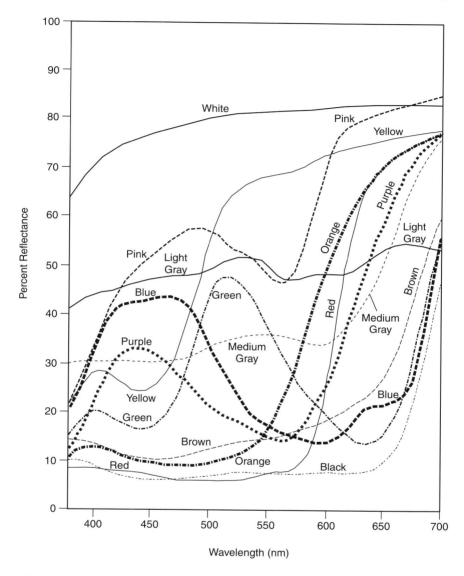

FIGURE 5   Spectral reflectance curves of ordinary colored construction paper. Percent diffuse reflectance is plotted against wavelength.   The color names are those given by experimental subjects (from Boynton, 1990).

or a "blue" light on these same paper samples and the absolute values of the light reflected will vary enormously. In most natural settings, of course, objects are not illuminated by diffuse white light nor are they uniformly pigmented. So the "shape" of the light reflected from a natural object is both complex and ever changing.

Finally, objects do not sit in isolation, one from the other. When, for example, one walks in the woods, light is both filtered through and reflected from the canopy of leaves overhead and the bushes around, shifting the SPD of the ambient light significantly toward the medium wavelength "green" part of the spectrum. (Think here of photographs taken inside a tent—say, a green tent—and the spectral shift becomes more intuitive.) Moreover, when two objects sit side by side, light reflected from one will be cast upon the other, such that each object acts, in effect, as a local filter for the light source. A blue teapot sitting on a white table gives the table a blue tinge; the terra-cotta red floor casts red light on the blue kitchen chairs. Despite the pervasive nature and significant effects of the interreflection (by one estimate, in a natural scene, fully 15 percent of an object's reflected light is contributed by its nearby objects) (Funt and Cardei, 2000), it is not one that we are able to see in ordinary color vision. Indeed, most of the above facts are "hidden" from us by our visual system.

The spectral world, then, is a complex one, and it is in this world that all vision—human vision included—evolved. The general evolutionary "question" posed to each and every developing visual system, then, was not "what use could a visual system make of *color*?" Rather, the "question" posed was "what use could a visual system make of *spectral information*?" What could a visual system do with wavelength, as opposed to luminance?

Two Principles of Luminance Vision

In answering this question, I believe that there are two widely accepted general principles of luminance vision that might be equally applicable to spectral vision—two principles that can help us understand "color" vision in ourselves and other species as well.

Perhaps the most basic principle of vision is simply this: *The proximal stimulus of all vision is luminance contrast*, not light per se or absolute luminance. To "see" anything at all, be it only the detection of a large looming shadow, there must be contrast in the light stimulus. To put this another way, if one wants to see what is in a picture, a black object against a black background will not prove very informative. What one needs is a dark figure against a light background—or, conversely, a light figure against a dark background. What is essential to vision are not the absolute levels of light per se, but luminance contrast in the image, be it positive or negative. Thus, the most important of visual information is where and how much contrast exists in the image.

The second most basic principle is as follows: *The more simple the behavioral repertoire of the organism, the more specific the information encoded by the initial sensory processes; the more complicated the behavioral repertoire of the creature, the more general the information encoded* (Lennie et al., 1990). To see what this principle involves, begin with the frog's four classes of retinal ganglion cells (Lettvin et al., 1959) (ganglion cells are the "output" cells from the retina). One kind of cell is sensitive to borders between light and dark areas, another to moving edges, another to reductions in overall illumination, and the last to small dark moving convex edges. The information processed by each ganglion cell of the frog is considered "specific" because of the discrepancy between the complex properties of the image that fall on the frog's retina and the limited number of those properties to which the ganglion cells react (i.e., because of the loss of information due to neural filtering). For example, sit a frog in front of the Mona Lisa, and the image falling on its retinae will form a complex pattern of spectral and luminance contrasts—the same pattern that the image would fall on your retinae if you were looking at the Mona Lisa. But because the ganglion cells of the frog respond to only four distinct patterns of illumination, there will be little ganglion response—probably only the "border detectors" will fire, responding to the frame of the picture. In this way, the frog's ganglion cells miss or serve to filter out most of the properties of the proximal stimulus, the complex image. (This is also why frogs make lousy art critics.)

For the frog, of course, such highly filtered information is behaviorally appropriate. As Lettvin et al. (1959) first claimed, in the natural environment of the frog, these light patterns on the frog's retina are usually caused by particular kinds of events in the world: dimming illumination over a substantial portion of the retina is caused by the movement of a largish opaque object, an object that is often a predator; small convex moving edges are usually caused by flying insects; borders between light and dark areas are usually caused by the borders of objects; and moving edges are usually caused by moving objects (or the frog's own movement). Thus, one finds a more or less direct wiring from these classes of cells to pathways that control the frog's behavior in relevant ways: The ganglion cells that react to looming shadows stimulate evasive behavior; those that are sensitive to small dark moving spots cause tongue-swiping behavior and so on.

In a mammalian or primate visual system, the initial encoding of the retinal image must subserve an indefinite number of complex behaviors. As Homo sapiens, we can walk through a tangled forest floor, thread a needle, recognize an acquaintance in a concert crowd, pick green peas, park a car, dodge (or catch) a speeding missile, secure a squirming baby into a pair of overalls, read fine newsprint, or swat a passing fly. Such diversity of behavior is produced by equally diverse computational tasks, and these in turn have diverse informational requirements. However information about the retinal image is encoded, as much information as possible about the image itself must be retained (not for us, mere edge detection for the Mona Lisa).

What then is the most general form of visual encoding? Return once again to the first principle of vision. If the proximal stimulus of vision is luminance contrast, then what primate vision must encode is where and how much luminance contrast exists in the retinal image, hence the spatially opponent center-surround organization of retinal cells in all vertebrates that encode fine-grained contrast information about the retinal image. Each center-surround cell "watches" a particular circular area of visual space, an area that is divided by the cell's response into a "center" (a smaller circular area in the middle) and a "surround" (the outer concentric region). Here there are two basic classes of cells, "ON" and "OFF." When a bright light is shone in the center region of an ON-center cell, it fires in response to the illumination; when light is shone on the surround, such a cell is inhibited or prevented from responding. An "OFF-center" cell has just the opposite response (i.e., it is excited by light in its surround but inhibited by light in the center region). Thus the response of a center-surround cell corresponds to either positive or negative luminance contrast between the center and the surround regions within a particular area of visual space (its "visual field"). In primates, the ON and OFF cells form two separate systems, from the retina to primary visual cortex, that are anatomically and physically distinct (Schiller, 1982, 1984; Schiller et al., 1986). This separation extends the range of contrast signaling (each cell has to signal only the positive range or the negative range) and makes for a "speedier," less noisy system (by using an excitatory signal for the transmission of all luminance information).

It seems to me that these same two principles might well apply to what I will call spectral vision, namely, that

1. Given a primate's large repertoire of complex, visually guided behaviors, the primate visual system requires a general encoding of retinal information.
2. What is encoded at the retina is contrast information—both luminance and spectral contrast information—that is then used, both separately and conjointly, to solve the multiple computational problems of primate vision.

Explaining this will take a good deal of work. As is commonly acknowledged, creatures with one kind of cone (or daylight retinal receptor) are color blind: They are unable to discriminate stimuli on the basis of wavelength alone. Although the receptor reacts to light across a broad spectrum of wavelengths, by increasing or decreasing the intensity of the light stimulus, the receptor's response can also be increased or decreased (Figure 6). Thus a particular response does not signal the presence of any particular wavelength: Given a single receptor, intensity and wavelength are conflated. Nonetheless, even creatures with only one kind of receptor could make use of spectral information. Indeed, one way to think of retinal receptors, given their preferential response to a small range of wavelengths (e.g., the "red" range), is as a spectral filter. Consider the black and white photo in Figure 7. Here, variations along the gray scale signal the intensity differences in the visual scene. When a color filter is added to the camera lens, however, the luminance values change. For example, when a blue filter (a filter through which blue light is transmitted but

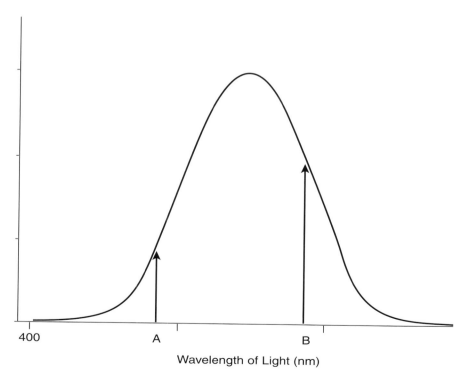

400   A   B

Wavelength of Light (nm)

FIGURE 6   A one-pigment system (from Sekuler and Blake, 1990).

FIGURE 7   A black and white photograph taken without any color filters.

which filters out red light) is attached to the camera lens, the apples in the scene become very dark, almost black (Figure 8). Indeed, almost all the fruit in this scene are somewhat darkened, a fact that tells us that most of the fruit reflect a reasonable amount of long wavelength or "red" light. In a natural setting, a spectral filter on a "black and white" (or single-cone system) can be extremely useful in enhancing luminance contrast. For example, if one were a creature whose diet consisted mainly of ripe red fruit set against a background of green leaves, a permanent blue filter of just this sort (i.e., a single retinal receptor insensitive to red light) would prove very useful.

In fact, this visual "strategy" of using a receptor as a filter to enhance luminance contrast is found in many species. For example, Lythgoe (1979) hypothesized that the rods and cones in certain fish have evolved to function as filters that highlight the contours of objects against the background space light. If the photo pigment is "matched" (maximally sensitive) to the spectral range of the background light (i.e., if the filter is matched to the "color" of the water), then a dark object will be highlighted against the brighter background; if the photo pigment is "offset" from the dominant wavelengths of the background light, then a bright object will be outlined against a poorly illuminated background. The skipjack tuna, then, which spots its prey from below (a dark object against a bright surface), has only

FIGURE 8   The same scene as depicted in Figure 7 using a blue filter. Note how dark the red fruits and vegetables appear, whereas the blue colander is now light gray.

one photo pigment that is matched to the background light. The walleye, bluegill, and piranha that inhabit dark, particle-laden (hence, red-shifted) waters are all dichromats (i.e., have two types of cones), with one cone type matched to the near-infrared, a wavelength common to their "black" water habitat during the dusk and dawn hours of feeding (Levine and MacNichol, 1979; Lythgoe, 1979; Munz and MacFarland, 1977).

Apart from spectral filtering, how might spectral information be used? In particular, what widespread benefit might the addition of yet another, differently tuned receptor bring about? In the standard version, what an organism gains through the addition of another receptor type is the ability to do "rough" wavelength comparison (Figure 9). Given a light stimulus of a single wavelength, one can compare the response of receptor 1 with the response of receptor 2 to determine the unique wavelength of the stimulus, a ratio that will remain constant independent of any change in intensity. Unfortunately, adding another cone yields only "rough" wavelength discrimination because any dichromat can be "fooled" by a light of two or more wavelengths: For any given ratio of cone response caused by a single-wavelength light, it is possible to produce that same ratio using a stimulus of two or more wavelengths (Figure 10) Thus, it is not possible for the system to disambiguate the two types of stimuli. By adding more receptors, a system can become capable of more accurate wavelength dis-

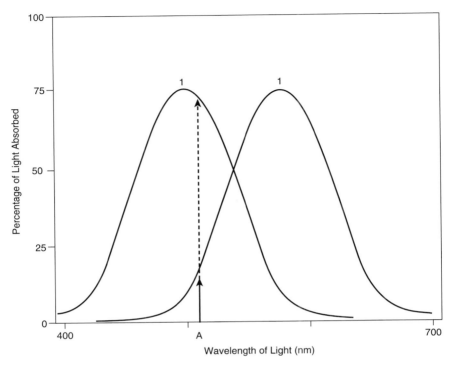

FIGURE 9   In a two-pigment system, the responses of receptors 1 and 2 can be compared to determine the unique wavelength of a stimulus (from Sekuler and Blake, 1990).

crimination, but the same principle holds no matter how many different types of cones are added. (A three-cone system can be fooled by a stimulus composed of three lights, a four-cone system can be fooled by a stimulus of four lights, and so on.) Again, according to common wisdom, the number of cones that a color system will have is constrained by the system's particular need for spatial resolution (the more cones there are, the less the spatial resolution of the system) as well as the constraints imposed by the (natural) spectral environment.

It is an interesting fact, however, that the vast majority of mammals, all of which have sophisticated visual systems, are dichromats. So, in the standard version, what each mammalian species has gained is only very rough wavelength discrimination (i.e., bad color vision). So why, exactly, would that ability have been so uniformly useful? For the version of spectral vision that I put forward here, one can think of two differently tuned

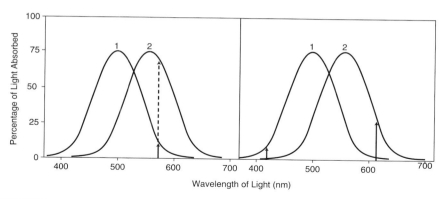

FIGURE 10 In a two-pigment system, the same ratio of response that is produced by a stimulus of a single wavelength (left), can be produced by a stimulus composed of two (or more) different wavelengths (right) (from Sekuler and Blake, 1990).

receptors as creating a means to compare two *ranges* of wavelengths (of comparing two ends of the spectrum) divided at that point at which both receptors give the same response (Figure 11). Let receptor 1 be given a positive signal and receptor 2 be given a negative one. In this antagonistic relation, a positive cumulative response will signal a proximal stimulus that has wavelengths predominantly in the lower range of the spectrum whereas a negative response will signal a proximal stimulus with wavelengths predominantly in the upper range. What will be signaled in a spectral system, in other words, are "positive" and "negative" ranges of a spectrum, just as in the luminance system, ON and OFF cells signal the positive and negative ranges of luminance.

In very primitive visual systems, we can see this kind of spectral contrast used for primitive motor tasks. Take for example the copepod *Daphnia* (Menzel, 1979). In blue light, *Daphnia* become highly active, lean forward, and move steadily at right angles to the angle of illumination; in red light, *Daphnia* remain upright and move slowly in parallel with the angle of illumination. One hypothesis is that this is related to foraging behavior. In phytoplankton-rich layers of water, where the chlorophyll of the plankton absorbs blue light and hence produces red ambient light, *Daphnia* slow down to eat; in phytoplankton-poor layers, in which the water is more blue, *Daphnia* begin to forage. In this way, *Daphnia* exhibit a typical pattern of wavelength-specific behavior, in which blue and red light produce positive and negative phototaxis, respectively.

In more complex visual systems, spectral contrast, of and by itself without spatial contrast, would not be of much use in discerning distal

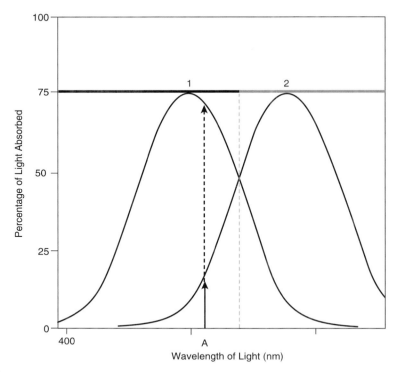

FIGURE 11   Here, the two receptors are placed in an antagonistic relation: receptor 1 provides a positive input and receptor 2 provides a negative input. When
the combined signal is positive, the wavelengths of the light source will be predominantly from the blue range of wavelengths; when there is a negative response,
the wavelengths of the light source will be predominantly from the red range of
wavelengths (adapted from Sekuler and Blake, 1990).

objects and their properties. In the primate visual system, this problem
seems to be solved by "interweaving" the encoding of luminance and
spectral contrast. In our own case, we have three types of daylight receptors—cones with overlapping ranges of response. Each cone responds to
a broad range of wavelengths, each with a different peak sensitivity (Figure 12). Thus, they are called the blue, green, and red cones or, more
properly, the short (S), medium (M), and long (L) cones, respectively. At
the level of retinal ganglion cells, 52 percent of all ganglion cells are color-
opponent cells, with the same spatially opponent center-surround organization as luminance cells (which receive input from the same kind of
cone). As one can see, the majority of cells are red-green opponent cells—
center–surround cells with an L ON-center and an M OFF-surround (L-M

cells) or the reverse arrangement, M ON-center and L OFF-surround (M-L). Blue-yellow cells, S ON and OFF center cells, comprise only 7 percent of all ganglion cells (Malpeli and Schiller, 1978; Zrenner and Gouras, 1981). Exactly what, for example, an L-M ganglion cell signals is an interesting question. Contrary to intuition, it does not signal spectral contrast between the center and the surround per se. It is true that such a cell will respond to red light in the center region and green light in the surround, but it will also respond to, for example, a red light shone only in the center, to a diffuse red light over the entire center–surround area, or to a lone white light in the center. (However, such a cell does respond uniquely to spectral contrast across a moving border, in virtue of the pattern of inhibition and excitation as the border passes first through the surround, then the center, and then the surround again.) So L-M center–surround cells are informationally complex, carrying interwoven information about spectral and luminance contrast.

This view that the color and luminance systems both serve to encode "visual" contrast information makes sense of the kinds of sensitivity control that exist at the retina and early in cortical processing. Start with the

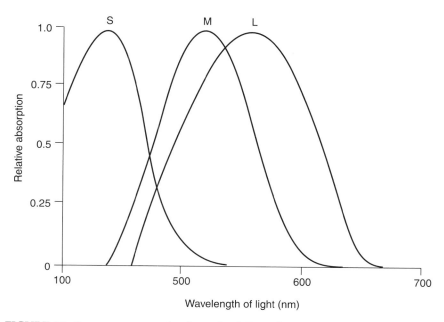

FIGURE 12 Response properties from the S, L, and M cones of the human retina (from Sekuler and Blake, 1990).

luminance system. In luminance processing, the visual system must operate across the very broad range of luminance throughout the day—nearly 10 decades from noon sunlight to starlight (Walraven et al., 1990)—yet a sensory neuron has a maximal firing rate of several hundred spikes per second. The range of stimulation over which the system must respond far exceeds the ability of a sensory neuron to signal fine-grained distinctions. For the most part, when this problem of visual sensitivity is discussed, it is framed as one of *brightness constancy*. How does the visual system ensure that an object is seen as having the same brightness (lightness) despite changes in luminance, that a piece of white newsprint looks white both outdoors and in? For example, Fiorentini et al. (1990) quote Hering (1878) to illustrate the problem: "without this approximate constancy, a piece of chalk on a cloudy day would manifest the same color as a piece of coal does on a sunny day, and in the course of a single day it would have to assume all possible colors that lie between black and white." This is true but perhaps not the best way to see the problem. If the visual system responds to contrast as opposed to the absolute luminance values, then the problem posed by variable illumination is not one of object reflectance per se but of *contrast sensitivity*: The central problem is not whether paper or chalk continues to look white throughout the day but whether one can find the chalk on a white table at high noon or read the black newsprint on white paper in dim light. What is important is the relative luminance of an object and its background (i.e., *contrast*). Theoretically, that is, a single neuron could respond across the entire range of daily illumination but only at the expense of fine-grained contrast resolution. Each small increase in firing rate would signal a large increase in illumination, and small contrast changes in the image would yield no response. Conversely, if the luminance range of the neuron were narrowly restricted (say, to the noon sun), the neuron would be sensitive to contrast within that range but otherwise "blind." Somehow, the system must "discount the illuminant" and signal contrast with a constant signal. Note that if a global solution to contrast sensitivity were implemented, then something close to "brightness constancy" would also result.

One can think of the spectral system as facing a very similar problem. Here, the primate visual system must respond to *spectral contrast* in the face of constant changes in the SPD of sunlight, both throughout the day as the spectral composition of sunlight changes, but also as we move through the environment, turn toward and away from the sun, walk under the shade of trees, and so on. Just as in luminance processing, the problem is not that of color constancy per se, of trying to ensure that a yellow banana looks yellow from dawn to dusk, but of seeing the ripe banana at all, of "segmenting" the ripe banana quickly from its green counterparts. In a species with a very limited behavioral repertoire or for one that lives in slowly changing and/or predictable "spectral worlds," there is an easy way to maintain or even enhance spectral contrast given a change in SPD: Simply change the spectral filters. To return to the fish again, the action spectra of the cones of the Pacific salmon, which migrate between fresh water and the ocean, change seasonally so as to maintain optimal contrast between prey and the different types of water (Munz and MacFarland, 1977; Whitmore and Bowmaker, 1989). For us, however, there is no simple solution of this sort: We are interested in a vast number of differently colored objects, against multiple backgrounds, under conditions of constantly changing SPDs. What, then, ought the visual system to do? On the assumption that there are no spectral biases that span across all these variables, the system's best global strategy for discerning spectral contrast is to "discount" the predominant wavelength in an image should there be one. All things being equal, viewing the visual scene as if under a white light will not enhance particular contrasts per se, but, for the most part, it will allow us to see whatever spectral contrasts between objects and their backgrounds actually exist. For example, if the object of interest is a bright blue, and the SPD of the illuminant is shifted toward the long wavelength end of the spectrum, toward the "reds," the blue object's natural contrast with its red background will be obscured (i.e., very little blue light will be reflected). Under conditions of an uneven SPD, such adaptation produces a rough approximation to color constancy because the spectral biases of the actual illuminant will be discounted. Thus, when one stands with a yellow banana in hand, the banana will look (more or less) yellow whether you are facing toward or away from the sun, given retinal adaptation.

Note that this kind of retinal adaptation does not distinguish between an illuminant with an uneven SPD and what I call "the bordello effect," when the objects in the visual scene all have a common hue, say, red. Here, retinal adaptation will normally *increase* the natural spectral contrast between objects and their backgrounds. If, for example, the visual system is trying to segment red velvet curtains against red-flocked wallpaper, filtering out the common "red" wavelengths may well leave differing residual spectral profiles if, for example, the curtains are a little more orange than the wallpaper. These are precisely the conditions in which color constancy fails. Being able to distinguish the curtains from the wallpaper is more important than seeing the curtains as being the correct shade of red.

To summarize, the initial levels of visual processing (in humans) serve to encode "general" contrast information—information about both luminance and spectral contrast. I suggest that this contrast information is used both conjointly and separately by visual cortex to solve the many problems of vision—the problems of discerning, among other properties, the position, shape, depth, and motion of objects within the visual scene. Just as the luminance system does not function in order to see brightness and darkness per se, neither does the spectral system function for the sole purpose of seeing the colors. Both the spectral and the luminance systems function *to see objects simpliciter*. Of course, we do see objects as being brighter or darker, and we do see objects and other media as colored, so doing so must be *one* of the functions of the luminance and spectral systems, respectively. But what looks from our end, as conscious perceivers, as the raison d'être of human color vision (i.e., seeing the world "in color") may not be its primary function at all. We see the world in color, in other words, not because *the colors* have proven so very useful but because spectral contrast information has proven so.

Benefits of a Spectral System

Unfortunately, the above view of the spectral and luminance systems has rarely been considered seriously for one simple reason: For any given visual image, wherever there is *spectral* information in an image, there is usually *luminance* information present as well. That is, if one were to take a black and white photo of, say, a patch of strawberry plants, one would find that red berries differ in "brightness" from both the green berries and from the green leaves as well. So even without spectral information, it is possible to spot the strawberries against the leaves and to tell the ripe from the unripe berries. But if it is possible to perform almost all of the requisite visual tasks using luminance information alone, what benefit would adding a spectral system confer? Spectral information would seem redundant.

The short answer to this question is this: "Pure possibility"—the range of computable functions theoretically available to an evolving system—is perhaps the least of the constraints acting upon any visual system as complex as primate vision. In evolution, our visual system has been shaped by many requirements, three of which are as follows.

First, although we are often told how large the human brain is, our brains are *not* infinite, and relative to the complexity of what they manage to do, perhaps we ought not to think of them as very large at all. In the face of finite resources, "cheap" computational solutions are preferable to those which use significantly greater computational resources. If, by using spectral information, there is a simpler way to solve a visual processing problem, all things being equal, spectral processing might be a preferable method. Going back to the strawberries again, although it is possible to pick strawberries using luminance information alone, doing so would be a significantly noisy process given the paucity of such information. But if the input from "red" cones were used as a spectral filter to increase the contrast between the ripe strawberries (and only the ripe strawberries) and the background green foliage, noise would be reduced, and the need for costly noise reduction mechanisms would be eliminated. Sometimes spectral information ought to provide an easier means of performing the same task.

Second, the real challenge facing primate vision is not task performance per se, of discerning this or that property of the visual environment, but of real-time processing. To make a visual system that can recognize a flight of stairs is a reasonably complex task. But to make a system that will recognize that flight of steps before you take the fatal step is a far more challenging affair. Real-time visual processing—having a visual system that can guide our movement at a reasonably rapid pace—is one of the great design hurdles which evolution has had to overcome. Hence, any system that can provide a faster solution to a visual processing problem has enormous utility.

Finally, primates are soft and squishy creatures moving about in a relatively dangerous environment. Hence, for us at least, evolution has placed a high premium on the reliability of our perceptual systems. You cannot afford to recognize that flight of stairs only 99 percent of the time, or to misjudge, one time out of a hundred, the distance from one step to another. In this context at least, 99 percent accuracy in a visual system would be exceedingly costly to life and limb. One standard tactic for increasing reliability is to compare "your" answer with someone else's—or, in this case, to compare an answer based on luminance information with an answer calculated from spectral information. Given two different sources of information, and potentially two different means of solving a common problem, the reliability of the luminance system could be significantly increased. To put this another way, although, prima facie, computational redundancy might seem like a bad idea for any computational system—a squandering of resources—this is not so when the computational problems are difficult and the cost of failure is high.

In light of the above three constraints, when we examine experimentally the computational processes of primate vision, *we should see spectral contrast information used in parallel and complimentary ways to luminance information.* Sometimes visual processing will benefit from the independent processing of spectral and luminance information (in order to compare the answers); sometimes it will benefit from a single mechanism that can use both spectral and luminance information (in order to increase the available contrast information); and sometimes task-specific mechanisms, using only spectral or luminance information, will be more appropriate. All of this will depend on the exact task at hand and on the features of the visual scene. More specifically, I expect that spectral information will be utilized when

(a) Luminance information is simply missing from the retinal image. As I stated above, in the natural world, isoluminance is a very rare problem: Even the red (ripe) berries and the green (unripe) berries are distinguishable, in principle, on the basis of luminance measures. Nonetheless, crucial luminance contrast information is often lost locally. For example, in the black and white photograph in Figure 13, it is difficult to make out the features of the two figures on the bed because the shadows obscure the crucial borders between the faces and the background bedding. What is surprising about this photo is that it was taken in very bright morning light, conditions under which the normal observer would see the two figures quite easily. (In other words, what the photographer realized was that, with all spectral information expunged, the scene would have a mysterious and "veiled" quality created by the shadowing.)

(b) Luminance information is ambiguous between two incompatible answers. For example, in a visual image that contains only luminance information, dark moving spots across a retinal image can be caused either by a moving object (e.g., a leopard moving through the grass) or by moving shadows across a stationary object (as when the wind moves

FIGURE 13   In this photograph, which was taken in bright, early morning sun, shadows obscure many of the crucial contours of the two sleeping subjects. Photograph by Gena Hahn.

leaves back and forth and dappled light plays across the scene). However, when the leopard moves, his spectral contrast goes with him; when shadows move across a stationary object, there is no accompanying movement of spectral contrast. Thus, the spectral contrast information can disambiguate the scene.

(c) Luminance information is limited or noisy. For example, if you were asked to pick out the blue jellies in the photo in Figure 14, it is certainly possible to do so based on luminance information alone. Given the color photo in Figure 15, however, it is clearly faster and easier (for us) to use spectral information as well. When an object and its neighbors have low luminance contrast but high spectral contrast, a visual search using both spectral and luminance contrast information might be faster and cheaper.

Recent experiments show that, indeed, spectral information contributes to some aspect of virtually all human visual abilities including those that previous research had dubbed "color blind." The research on motion processing provides a good example here.

In 1978, Ramachandran and Gregory published the results of the first isoluminant motion studies. The tests used two random dot arrays, identical except that in one array a central square-shaped region of dots was shifted horizontally. When superimposed optically, and presented alternately, the display produced the impression of a square oscillating back and forth. Under isoluminant conditions, using a red-green random dot array, however, there was no apparent motion. Later experiments suggested that the color contribution to motion processing was either absent (Livingstone and Hubel, 1984, 1987, 1988; Ramachandran, 1987) or very weak in comparison with that of luminance (Derrington and Badcock, 1985; Mullen, 1985; Mullen and Baker, 1985). Most strikingly, Cavanagh et al. (1984) found that certain isoluminant apparent motion stimuli (a red-green isoluminant grating phase shifted left or right) appeared either *stationary* or to be moving *very slowly*.

In the early 1990s, however, Papathomas et al. (1991) asked the following question: even if color is neither necessary nor sufficient for apparent motion perception, might color still not contribute significantly to its processing? In their experiments they disassociated color and luminance cues in an apparent motion display, asking what would happen when (a) the luminance cues were ambiguous but the color stimuli were directional (i.e., were all shifted in a particular direction); (b) when both color and luminance stimuli were directional and coordinated (i.e., shifted in the same direction); and (c) when the color and luminance cues were directional but conflicted (i.e., shifted in opposite directions). As suspected, they found that adding color information resolved ambiguity in an isoluminant display and also enhanced motion perception when

FIGURE 14 In this photograph one can see that each color of jelly differs from the other colors in brightness; hence, in principle, one could select a raspberry (red) jelly by luminance information alone (from Solomon, 1991).

FIGURE 15 Given the addition of spectral information, it is much easier for us to discern the raspberry (red) from the lime (green) jellies.

added to a directional isoluminant display. Most interesting, in (c), color was able to "capture" the direction of motion given low luminance: As luminance decreased, there was a point, different for each observer, at which color, not luminance, determined the direction of motion. When luminance information is slight or noisy, the visual system used color information.

Similarly, Mullen and Boulton (1992) investigated Cavanagh et al.'s (1984) reports that isoluminant motion displays were perceived as either slowly moving or stationary. They divided the task into three components: the detection of motion, the identification of its direction (i.e., left or right), and the perception of velocity or smooth motion. Mullen and Boulton found that, for isoluminant stimuli, subjects could both detect and identify motion at near-threshold spectral contrast, over a wide range of speeds and grating widths. In fact, for these two tasks, detection and identification, spectral processing proved about on par with luminance processing. The perception of smooth motion, however, was possible only under very specific conditions—for gratings with very high spectral contrast and moving at slow speeds. Thus, at low spectral contrasts, subjects found themselves in the odd position of detecting and identifying the motion yet without the *perception* of that motion *as* smooth motion. These findings were both confirmed and refined by Derrington and Henning (1993). In their experiments they found that, for gratings moving at very slow speeds (below 2 Hz), the chromatic system was actually slightly better at motion identification, whereas the luminance system was slightly better at detecting motion in the parafoveal areas.

Based on these results and a variety of psychophysical experiments (Gorea et al., 1993; Thompson, 1982), Hawken et al. (1994) hypothesized that motion processing might have two separate components—mechanisms for fast- and slow-moving stimuli, that might be further subdivided between chromatic and luminance mechanisms. What they looked for in these experiments were distinct response characteristics for each of the four conditions: slow chromatic, slow luminance, fast chromatic, and fast luminance. They asked, for example, how do our visual abilities for slow-moving chromatic stimuli compare to our visual abilities for fast-moving chromatic stimuli? Again, subjects were tested for three motion tasks: motion detection, direction discrimination, and velocity perception. In velocity processing, the results showed the expected divide. At slow speeds, the perception of velocity was highly contrast dependent for both luminance and chromatic processing, whereas at high speeds (greater than 4 Hz) velocity perception, for either luminance or chromatic stimuli, showed very little contrast dependence. There was also a division, at slow speeds, between luminance and chromatic processing: Although both were highly contrast dependent, their contrast gains differed. That is, as slow velocities increased, it took a far greater increase in contrast to see the chromatic stimuli as moving smoothly than it did to see luminance stimuli moving smoothly, suggesting different mechanisms for luminance and chromatic stimuli. The experiments for motion detection and identification showed similar results (Gegenfurtner and Hawken, 1995): There was a marked difference between fast- and slow-motion processing, as well as a difference between luminance and chromatic processing for slow speeds. Hence, the authors concluded that there were three separate channels for motion processing—two separate channels for slow speeds, one chromatic and one luminance, but only one channel for high speeds, sensitive to both kinds of information. Finally, as a last twist in the story, Cropper and Derrington (1996), using a new masking paradigm, demonstrated that there is a color-specific motion mechanism for analyzing the direction of motion for very fast or brief signals, thereby confirming the most controversial results of Gegenfurtner and Hawkins.

The results from recent experiments on depth perception have followed much the same pattern. For example, Kovacs and Julesz (1992) tested subjects for the perception of depth using random dot stereograms. When polarity-reversed luminance stimuli were used, the subjects were unable to perceive depth in the stereoscopic displays from luminance information alone (i.e., the displaced central square of dots did not appear as raised relative to the background), but could do so when chromatic information was added (i.e., the square "popped out"). Just as in the Papathomas et al. (1991) experiments on apparent motion, the chromatic information provided the requisite additional "resolving power" in this particular depth perception task. Tyler and Cavanagh (1991) have also reported that a strong percept of stereoscopic motion can be produced by chromatic stimuli alone, and that given the differences in thresholds for luminance and chromatic stimuli, these stereoscopic abilities might well be supported by different neural pathways. In addition, Zimmerman et al. (1995) have reported that depth perception from purely pictorial cues is possible under equiluminant conditions. Taken together, these and other results (Dengler and Nitschke, 1993; Faubert, 1995; Jimenez et al., 1997; Kingdom and Simmons, 1996; Kingdom et al., 1999; Simmons and Kingdom, 1995, 1997; Wanatabe and Cavanagh, 1992) suggest that spectral information is used by many of the mechanisms of depth perception.

In addition, spectral information seems to be used for texture segregation (Gorea and Papathomas, 1991; Pessoa et al., 1996), for maintaining coherence in the perception of moving objects (Cropper et al., 1996; Kooi and De Valois, 1992), and attention capture (Baylis and Driver, 1992; Folk et al., 1994; Motter, 1994; Nagy and Cone, 1996; Nothdurft, 1993; Theeuwes, 1992). The spectral system, it would seem, provides information, in many guises, for a large variety of visual tasks.

Cerebral Achromatopsia Revisited: Philosophical Conclusions

Let us return to the case of M.S., the cerebral achromatopsic patient discussed above. One puzzle presented by his case was a neurophysiological one. As Heywood et al. (1994) described the case, even with bilateral damage to the cortical "color center," M.S. retains some color abilities but not others. Hence, the concluding question of their paper is: What "residual" color information is available and where is it available in M.S.'s visual system?

If something like the above view of color vision is correct, then this way of framing the question may well misconstrue the problem. As suggested above, to assume that there must be a color area is to assume that what we see consciously provides the raison d'être of the spectral system—that the purpose of the color system is to produce "the colors." Thus, spectral processing must eventuate in a place where the colors are represented and that, thereafter, provides information about the colors (so encoded) to other visual areas. But if this were true, M.S.'s behavior seems truly mysterious. After all, M.S. can pick out a large red square under circumstances in which he cannot pick out a small red square; he can also pick out a large red cross on a random luminance checkerboard, but he cannot pick out the large red numbers in Ishihara pseudo-isochromatic plates. If M.S.'s representation of our phenomenal color space is selectively damaged, then why are his perceptions of, say, all red figures not selectively annihilated? Why is red available on some occasions but not on others (and the same, *mutatis mutandi*, for the other colors)?

On the view of color vision I gave above, seeing objects simpliciter is the evolutionary raison d'être of the spectral system, not seeing objects as colored; thus its primary function is to supply spectral contrast information for motion identification, pattern analysis, depth perception, texture discrimination, and the like. Looking at things in this way has the following theoretical consequences.

First, although bilateral cortical damage to temporo-occipital cortex has surely caused M.S. to loose numerous spectral capacities, most notably the ability to consciously discriminate and categorize hue, the abilities he retains are not the "residual" abilities of a damaged "color center" at all. Rather, when the spectral system is damaged, certain spectral mechanisms will either be damaged or lose their spectral (as opposed to color) input. Experimentally, one would expect a patient to show a wide variety of spectral deficits just as M.S. does. On the one hand, for visual tasks that are impossible without spectral contrast information, the patient ought to be able to perform some of these tasks but not others. For example, M.S. could see apparent motion in an experimental display (a task that requires signed spectral contrast) but could not do color matching. Here the precise nature of the task and, hence, of the stimuli will be of prime importance. For example, perhaps the salient difference between the concealed numerals in the Ishihara plates and the large forms (squares and crosses) on the luminance checkerboards is that the squares and crosses have continuous color borders, whereas the numbers are composed of individual spots. To see the numbers, the system must first see the dots composing the numerals as sharing a common property, be that property "greenness" or merely "positive spectral contrast," and then recognize that set of dots as having the shape of a numeral. Intuitively at least, this task would seem more difficult than whatever is involved in discerning continuous borders, a task that need not require "signed" contrast information at all. On the other hand, for visual tasks that do *not* by their very nature require spectral information such as the tasks of ordinary depth perception and motion detection, one might still see a number of deficits. Where spectral information is normally used to aid luminance processing, one might see reduced or slowed performance; under conditions of high noise or ambiguous luminance information, one might see a complete loss of function.

Second, there need not be a "color area" in the standard sense at all, a physiological area of the brain which serves to map the phenomenal space of color (Zeki, 1980, 1990), the coordinates of which determine the nature of our color phenomenology. If the function of both luminance and spectral vision is the same, then the possibility of such a color area is about as prima facie plausible as the existence of a "brightness" center. (Imagine here an entirely color-blind people who assumed that, because they see objects of varying brightness, that the luminance system must eventuate in a place where luminance is encoded on a "gray scale" and which then supplies this "gray scale" information to the "residual" functions of motion processing, etc.) To put this another way, one possibility not foreseen by Heywood et al. is that phenomenal color might ride on the coattails of other spectral processes, not the other way around. Perhaps whenever the visual system ascribes an "inherent" color to an object, the "colonial blue" of the kitchen chairs, it computes its answer "on the fly" using the spectral information already encoded for some other visual purpose. That is, perhaps the dimensions of phenomenal color space reflect the informational requirements—the spectral "packaging"—of these *other* visual tasks. This is certainly possible.

Finally, on the view of color that I am suggesting, determining the spectral reflectance profile of an object, its "inherent" determinate color, is but one specific task that the visual system performs with spectral information. One has to wonder just when and why the visual system would compute (something akin to) the spectral reflectance curve of an object given the difficulty of such a computation (Akins and Hahn, 2000).

Is it really the case, as our phenomenology leads us to believe, that each and every object is assigned a determinate spectral reflectance curve? Or do we merely see each object *as* having a determinate color, a color that could be roughly approximated were the visual system to turn its attention toward the task? Perhaps the visual system does not compute, as a matter of course, the spectral reflectance of each object within the visual field. Maybe it computes only rough color categories (e.g., "very bright red") or perhaps it makes do with only the spectral information already discerned for other spectral processes (e.g., "highly saturated color" or "positive spectral contrast"). That is, what we see on a given occasion might be a matter of how spectral information has been processed given that particular visual scene. Just as there may not be a "color area" that represents each determinate color, there may not be anywhere a representation of a determinate color each time an object is seen. This too is a possibility.

The second puzzle raised was a philosophical one about the nature of M.S.'s experience: What is it like to be M.S.? What is the nature of his visual experience? There are a number of philosophical lessons about color experience, both normal and pathological, to be gleaned, but here I will mention only two. First, one way out of the paradox presented above is to consider the possibility that there is some middle ground between seeing "in color" and seeing "in black and white." Objectively, of course, there either are or are not objects in the world with uneven spectral reflectance curves, like those of the colored construction paper in Figure 5; there either are or are not the sort of broadband sources of illumination that are necessary for color perception. None of this is in dispute. Rather, if seeing the chair as colonial blue is *not* the result of a "colonial blue" representation in the color center—if our ordinary perceptions of color are underwritten by any number of distinct spectral processes—then various aspects of our color phenomenology may be dissociable from each other. Theoretically, it ought to be possible to experience some spectral processes in the absence of others. Yet take away one or two "strands" of spectral processing—strands that need not map onto the dimensions of color we consciously recognize as aspects of color, such as lightness, hue, and saturation—and the resulting experience may not seem much like a color experience at all. Perhaps in the case of M.S. some but not all of the spectral processes that normally support our color phenomenology are absent. M.S. may not see the cross as any determinate color at all ("forest green"); he may not even see it as being in a certain color category ("green") at all. Nonetheless, M.S. may have some spectral information about the cross, some information that would contribute to a normal conscious perception of the cross as forest green. What he experiences in the daily run of things is some part (or parts) of our ordinary (nonpathological) color experience.

The problem is that M.S. does not *recognize* his present experience as such: He has no conscious access to that spectral information *as* some aspect of object color—hence his uninformative descriptions of the world as "dull," "washed out" or "gray."

Second, and more generally, this theory of color suggests that we, as normal subjects, may not be in the best position to determine the nature of our own color phenomenology. If, as I have just suggested, (a) we do not perceive the determinate color of each object as a matter of course, but only see each object as having some determinate color or other; or (b) if our color perceptions of what seems to be a single determinate color, "the very color of the kitchen chair, colonial blue," are the result of a number of dissociable processes; then (c) we too may fail to understand our own phenomenology. We may think we see each and every object with a determinate color; we may think that each color experience, or color "qualium," is a phenomenologically unified experience of a particular hue, "that every blue." But we may not see what we think we see. Just as M.S. may not be the best person to describe his own visual experience—or at least not without the help of a detailed theory of spectral processing that will allow him to conceptualize his conscious experiences of the world—we too may come to revise our understanding of our own experiences of a colored world.

REFERENCES

Akins, K., and M. Hahn. 2000. The peculiarity of color. In *Color: Vancouver Studies in Cognitive Science, Vol. 9.*, S. Davis, ed. Oxford: Oxford University Press.

Baylis, G., and J. Driver. 1992. Visual parsing and response competition: The effect of grouping factors. *Perception and Psychophysics* 51(2):145–162.

Berlin, B., and P. Kay. 1969. *Basic Color Terms: Their Universality and Evolution.* Berkeley: University of California Press.

Boynton, R. M. 1990. Human color perception. Pp. 211–253 in *Science of Vision*, K. N. Leibovic, ed. New York: Springer-Verlag.

Boynton, R., and C. Olson. 1987. Locating basic color terms in the OSA space. *Color Research Applications* 12: 94–105.

Burkhardt, D. 1989. UV vision: a bird's eye view of feathers. *Journal of Comparative Physiology* 164: 787–96.

Cavanagh, P., C. Tyler, and O. Favreau. 1984. Perceived velocity of moving chromatic gratings. *Journal of the Optical Society of America* 1:893–899.

Cropper, S., and A. Derrington. 1996. Rapid colour-specific detection of motion in human vision. *Nature* 379(6560):72–74.

Cropper, S., K. Mullen, and D. Badcock. 1996. Motion coherence across different chromatic axes. *Vision Research* 36(16): 2475–2488.

Damasio, A., T. Yamada, H. Damasio, J. Corbett, and J. McKee. 1980. Central achromatopsia: behavioural, anatomic, and physiologic aspects. *Neurology* 30:1064–1071.

Dengler, M., and W. Nitschke. 1993. Color stereopsis: A model for depth reversals based on border contrast. *Perception and Psychophysics* 53(2):150–156.

Derrington, A., and D. R. Babcock. 1985 The low level motion system has both chromatic and luminance inputs. *Vision Research* 25(12):1879–1884.

Derrington, A., and G. Henning. 1993. Detecting and discriminating the direction of motion of isoluminant inputs. *Vision Research* 33:799–811.

Faubert, J. 1995. Colour indiced stereopsis in images with achromatic information and only one other colour. *Vision Research* 35(22):3161–3167.

Fiorentini, A., G. Baumgartner, S. Magnussen, P. Schiller, and J. Thomas. 1990. The perception of brightness and darkness. In *Visual Perception*, J. Spillman and J. Werner, eds. San Diego, Calif.: Academic Press.

Folk, C., R. Remington, and J. Wright. 1994. Structure of attentional control: Contingent attentional capture by apparent motion, abrupt onset, and color. *Journal of Experimental Psychology: Human Perception and Performance* 20 (2):317–329.

Fuld, K., B. Wooten, and J. Whalen. 1981. The elemental hues of short-wave and extra-spectral lights. *Perception and Psychophysics* 29:317–322.

Funt, B., and V. Cardei. 2000. Computational uses of colour. In *Color: Vancouver Studies in Cognitive Science, Vol. 9.*, S. Davis, ed. Oxford: Oxford University Press.

Gegenfurtner, K., and M. Hawker. 1995. Temporal and chromatic properties of motion mechanisms. *Vision Research* 35(11):1547–1563.

Gorea, A., and T. Papathomas. 1991. Texture segregation by chromatic and achromatic visual pathways: an analogy with motion processing. *Journal of the Optical Society of America A* 8(2):386–393.

Gorea, A., T. V. Papathomas, and I. Kovacs. 1993. Motion perception with spatiotemporally matched chromatic and achromatic information reveals a "slow" and a "fast" motion system. *Vision Research* 33(17):2515–2534.

Hawken, M., K. Gegenfurtner, and C. Tang. 1994. Contrast dependence of colour and luminance motion mechanisms in human vision. *Nature* 367(6460):268–270.

Hendersen, S. 1977. *Daylight and Its Spectrum, 2nd ed.* Bristol, England: Adam Hilger.

Hering, E. 1878. *Zur Lehre von Lichtsinn* Wien: Gerald u. Sohne.

Heywood, C., A. Cowey, and F. Newcombe. 1994. On the role of parvocellular (P) and magnocellular (M) pathways in cerebral achromotopsia. *Brain* 117:245–254.

Jimenez, J., M. Rubino, E. Hita, and L. Jimenez del Barco. 1997. Influence of the luminance and opponent chromatic channels on stereopsis with random-dot stereograms. *Vision Research* 37(5):591–596.

Kelly, K., and D. Judd. 1976. *The ISCC-NBS method of designating colors and a dictionary of colour names.* NBS special publication 440. Washington, D.C.: U.S. Department of Commerce.

Kingdom, F., and D. Simmons. 1996. Stereoacuity and colour contrast. *Vision Research* 36(9):1311–1319.

Kingdom F. A., D. Simmons, and S. Rainville. 1999. On the apparent collapse of stereopsis in random-dot-stereograms at isoluminance. *Vision Research* 39(12):2127–2141.

Kooi, F., and K. De Valois. 1992. The role of color in the motion system. *Vision Research* (4):657–668.

Kovacs, I., and B. Julesz. 1992. Depth, motion, and static-flow perception at metaisoluminant color contrast. *Proceedings of the National Academy of Sciences of the United States of America* 89(21):10,390–10,394.

Lennie, P., C. Trevarthen, D. Van Essen, and H. Wassle. 1990. Parallel processing of visual information. In *Visual Perception*, J. Spillman and J. Werner, eds. San Diego, Calif.: Academic Press.

Lettvin, J., H. Matturana, W. Pitts, and W. McCulloch. 1959. What the frog's eye tells the frog's brain. *Proceedings of the Institute of Radio Engineers* 47:1940–1951.

Levine, J., and E. MacNichol. 1979. Visual pigments in teleost fishes: Effect of habitat, micro-habitat and behavior on visual system evolution. *Sensory Processes* 3:95–131.

Livingstone, M., and D. Hubel. 1984. Anatomy and physiology of a color system in the primate visual cortex. *Journal of Neuroscience* 4(1):309–356.

Livingstone, M., and D. Hubel. 1987. Connections between layer 4B of area 17 and the thick cytochrome oxidasestripes of area 18 in the squirrel monkey. *Journal of Neuroscience* 7(11):3371–3377.

Livingstone, M., and D. Hubel. 1988. Segregation of form, color, movement, and depth: Anatomy, physiology, and perception. *Science* 240:740–749.

Lythgoe, J. 1979. *The Ecology of Vision.* Oxford: Oxford University Press.

Malpeli, J., and P. Schiller. 1978. The lack of blue off-center cells in the visual system of the monkey. *Brain Research* 141:385–389.

Meadows, J. 1974. Disturbed perception of colours associated with localized cerebral lesions. *Brain* 97:615–632.

Menzel, R. 1979. Spectral sensitivity and color vision in invertebrates. In *Handbook of Sensory Physiology, Vol VII/6A: Comparative Physiology and Evolution of Vision in Vertebrates.* Berlin: Springer-Verlag.

Motter, B. 1994. Neural correlates of attentive selection for color or luminance in extrastriate area V4. *Journal of Neuroscience* 14(4):2178–2189.

Mullen, K. 1985. The contrast sensitivity of human color vision to red/green and blue/yellow chromatic gratings. *Journal of Physiology* 359:381–400.

Mullen, K., and C. Baker. 1985. A motion after affect from an isoluminant stimulus. *Vision Research* 25:685–688.

Mullen, K., and J. Boulton. 1992. Absence of smooth motion perception in color vision. *Vision Research* 32(3):483–488.

Munz, F., and F. MacFarland. 1977. Evolutionary adaptations of fishes to the photopic environment. In *Handbook of Sensory Physiology, Vol. VII/5: The Visual System of Vertebrates*, F. Crescitelli, ed. Berlin: Springer-Verlag.

Nagy, A., and S. Cone. 1996. Asymmetries in simple feature searches for color. *Vision Research* 36(18):2837–2847.

Nothdurft, H. 1993. The role of features in preattentive vision: Comparison of orientation, motion and color cues. *Vision Research* 33(14):1937–1958.

Papathomas, T., A. Gorea, and B. Julesz. 1991. Two carriers for motion perception: Color and luminance. *Vision Research* 31(11):1883–1892.

Pessoa, L., J. Beck, and E. Migolla. 1996. Perceived texture segregation in chromatic element-arrangement patterns: High-intensity interference. *Vision Research* 36(11):1687–1698.

Quine, W. V. 1960. *Word and Object.* Cambridge, Mass.: MIT Press.

Ramachandran, V. S. 1987. The interaction between color and motion in human vision. *Nature* 328(6131):645–647.

Ramachandran, V. S., and R. L. Gregory. 1978. Does colour provide an input to human motion perception? *Nature* 275(5675):55–56.

Rizzo, M., M. Nawrot, R. Blake, and A. Damasio. 1992. A human visual disorder resembling area V4 dysfunction in the monkey. *Neurology* 42:1175–1180.

Rizzo, M., V. Smith, J. Porkorny, and A. Damasio. 1993. Color perception profiles in central achromatopsia. *Neurology* 43:995–1001.

Schiller, P. H. 1982. Central connections on the retinal ON- and OFF-pathways. *Nature* 297:580–583.

Schiller, P. H. 1984. The connections of the retinal ON- and OFF-pathways on the lateral geniculate nucleus. *Vision Research* 24:923–932.

Schiller, P., J. Sandell, and J. Maunsell. 1986. Functions of the ON and OFF channels of the visual system. *Nature* 322:824–825.

Sekuler, R., and R. Blake. 1990. *Perception.* New York: McGraw-Hill.

Sellars, W. 1956. Empricism and the philosophy of mind. In *Minnesota Studies in the Philosophy of Science, Vol. 1*, H. Fiegel and M. Scriven eds. Minneapolis: University of Minnesota Press.

Shepard, R. N. 1997. The perceptual organization of colors: an adaptation to regularities of the terrestrial world? In *Readings on Color: the Science of Color, Vol. 2*, D. Hilbert and A. Bryne, eds. Cambridge, Mass.: MIT Press.

Simmons, D., and F. Kingdom. 1995. Differences between stereopsis with isoluminant and isochromatic stimuli. *Journal of the Optical Society of America A* 12(10):2094–2104.

Simmons, D., and Kingdom, F. 1997. On the independence of chormatic and achromatic stereopsis mechanisms. *Vision Research* 37(10):1271–1280.

Solomon, C. 1991. *Charmain Solomon's Thai Cookbook: A Complete Guide to the World's Most Exciting Cuisine*. Rutland,Vt.: Tuttle Press.

Sternheim, C., and R. Boynton. 1966. Uniqueness of perceived hues investigated with a continuous judgemental technique. *Journal of Experimental Psychology* 72:770–776.

Swain, M., and D. Ballard. 1991. Color indexing. *International Journal of Computer Vision* 7(1):11–32.

Theeuwees, J. 1992. Perceptual selectivity for color and form. *Perception and Psychophysics* 51(6):599–606.

Thompson, P. 1982. Perceived rate of movement depends upon contrast. *Vision Research* 22:377–380.

Tyler, C., and P. Cavanagh. 1991. Purely chromatic perception of motion in depth: Two eyes as sensitive as one. *Perception and Psychophysics* 49(1):53–61.

Walraven, J, C. Enroth-Cugell, D. Hood, D. MacLeod, and J. Schnapf. 1990. The control of visual sensitivity: Receptoral and postreceptoral processes. In *Visual Perception*, J. Spillman and J. Werner, eds. San Diego, Calif.: Academic Press.

Watanabe, T., and P. Cavanagh. 1992. Depth capture and transparency of regions bounded by illusory and chromatic contours. *Vision Research* 32(3):527–532.

Whitmore, A., and J. Bowmaker. 1989. Seasonal variation in cone sensitivity and short-wave absorbing visual pigments in the rudd, *Scardinius erythrophthalamus*. *Journal of Comparative Physiology A* 166:103–115.

Zeki, S. 1980. The representation of colours in the cerebral cortex. *Nature* 284:412–418.

Zeki, S. 1990. A century of cerebral achromatopsia. *Brain* 113: 1721–1777.

Zimmerman, G., G. Legge, and P. Cavanagh. 1995. Pictorial depth cues: A new slant. *Journal of the Optical Society of America A* 12(1):17–26.

Zrenner, E., and P. Gouras. 1981. Characteristics of the blue sensitive cone mechanism in primate retinal ganglion cells. *Vision Research* 21:1605–1609.

4

Why Neuroscience Needs Pioneers

Patricia Smith Churchland

As an academic, one shoulders an assortment of duties. These duties have different "hedonic scores," to put it in limbic system terms. Some of these duties involve grading marginally comprehensible and humorless term papers; some duties require that we hector brilliant but foot-dragging graduate students to cross the dissertation finish line. Yet other duties involve the tedium of the legendary department meeting, grinding on until the last dog has been hanged. By contrast, my duty here has a high hedonic score. So much so, that I hesitate to consider this brief commentary a *duty* at all.

From whatever direction we look at it, the awarding of the McDonnell Centennial Fellowships merits unalloyed glee. Each fellow has, after all, been awarded $1 million to pursue the intellectual adventure of his or her dreams. In addition, the award program is, I think, important for the particular flag it runs up the research flagpole. And that is the flag of the unconventional and unorthodox, the daring and risky, the not-yet gray, and the not yet medaled. An initiative such as the McDonnell Foundation Centennial Fellowship Program is important because, truly to thrive, science must nurture the intellectually adventurous.

Thomas Kuhn described science as predominantly paradigm governed, by which he meant that subfields have a kind of conventional wisdom that functions rather like a powerful gravitational well. The practitioners within the subfield tend to promote, and encourage, *and* fund those

Philosophy Department, University of California, San Diego

people whose projects and research essentially conform to the conventional wisdom prevailing in their subfield. Occasionally, against considerable odds, with considerable luck, and to considerable consternation, highly original ideas emerge to challenge the prevailing orthodoxy dominating a particular field of science. The pressures to acquiesce in the conventional wisdom, to work within it, further it, and confirm it, are, however, enormous. They exist at every level, from grammar school, college, graduate school, and all the way through the academic ranks. By and large, beautiful and brilliant science does indeed get done within the confines of the governing paradigm. Moreover, it is fair to say that progress would be hindered if kicking over the traces were the *norm*, rather than the "abnorm," as it were.

Making it *too* easy for the unconventional can, I have no doubt, result in a superabundance of cranks and crackpots—of visionaries with lots of vision but precious little common sense. Too much of a good thing may not, despite Mae West's winking assurances, be wonderful. Nevertheless, risky research is desirable if we are to unseat assumptions that are not so much proved as conventionally respected, not so much true as "*taken* to be true." When working assumptions go unquestioned, they tend to become dogma, however wrong they may be. What might have served well enough as a convenient fiction becomes, when canonized as *fact*, an obstacle to further progress. The trouble is, therefore, funding for the unorthodox researcher and his or her unorthodox hypotheses remains an ongoing challenge for the standard funding institutions, who by their very nature, rarely, if reasonably, support anything other than center-of-the-road projects.

Especially gratifying it is, therefore, that the McDonnell Centennial Fellowship Program was set up to reward the innovative and frisky who could undertake risky and surprising projects without kowtowing to the conventionally wise graybeards. It is all the more appropriate, given that the founder, James S. McDonnell, was himself a maverick, a man of genuine intellectual vision and courage. He knew very well the value of having the chance to develop an idea whose potential others were too timid or too rigid to see.

My recollection is that the inspiration for the Centennial Fellowship Program originated with Susan Fitzpatrick, and she deserves much credit for not only hatching the general idea, but for guiding it through the many delicate stages that brought us to this point, where we may rejoice with those chosen to receive McDonnell Centennial Fellowships. Many other people, including members of the McDonnell family and John Bruer, were of course crucial to the program's success, and we are in their debt.

I turn now to make several brief remarks about cognitive neuroscience in general. Were one to judge solely by what appears in the scientific journals and at scientific meetings, cognitive neuroscience is a research juggernaut plowing unswervingly and surefootedly forward. If you don't look too closely, it may seem that the basic framework for understanding how the brain works is essentially in place. It may seem that by and large most future neuroscience is pretty much a matter of filling in the details in a framework that is otherwise sturdy and established. But this appearance is largely an illusion.

Although more progress has been made in the past two decades in neuroscience than in all the rest of human history, neuroscience has yet to win its wings as a mature science. In this sense, neuroscience is rather like physics before Newton or chemistry before Dalton or molecular biology before Crick and Watson. That is, neuroscience cannot yet boast anything like an explanatory exoskeleton within which we can be reasonably confident that we are asking the right questions and using concepts that are empirically grounded and robustly defined. Molecular biology and cell biology, by contrast, do have considerable explanatory exoskeleton in place. Arguably, surprises still await us in these subfields, yet the general mechanisms governing the target phenomena are uncontestably better determined than in neuroscience.

This state of affairs does not mean that neuroscientists have been shiftless or remiss in some respect. Rather, it reflects the unavoidable problems inherent in understanding something as complex and conceptually alien as a nervous system. Moreover, neuroscience is likely to achieve maturity later than the other subfields, not just because it is harder— though it undoubtedly is—but also because it heavily depends on results in those other fields, as well as on technology made possible by very recent developments in physics and engineering.

When in grant-funding or paper-reviewing mode, we often have to pretend that the appearance of solidity is the reality. For sound pragmatic reasons, we pretend that there is a more or less well-established explanatory framework whose basic concepts are empirically well grounded and well understood. Hence we may assume, for example, that in nervous systems, learning mechanisms are known to be Hebbian, or that the spike is the only neuronal event that codes information. Allowed freely to reflect and ponder, we grudgingly suspect it isn't really so.

My students are shocked to be reminded that we do not in fact understand how neurons code information, or even how to distinguish neurons that code information from neurons that are doing something else, such as housekeeping or neuromodulation or something entirely "else." In fact, we do not have a concept of "information" that is suitable to what goes on in nervous systems. The Shannon–Weaver concept of information, designed for the context of communication lines, is not even roughly appropriate to our needs in neuroscience. "Information" and the "processing of information" are indispensable concepts in neuroscience, yet they lack theoretical and empirical infrastructure, much as the notion of "momentum" did before Newton or "gene" before 1953. Experiments make it clear that it is essential to talk about the brain as representing various things, such as the position of the limbs or a location in physical space, but we do not really understand what we need to *mean* by "representation." Nor do we understand how to integrate *representational* descriptions (e.g., "is a representation of a face") with *causal* descriptions (e.g., "is hungry").

This is only for starters. In addition, we do not understand how nervous systems exploit time to achieve cognitive and behavioral results. We do not understand how nervous systems integrate signals over time so that they can recognize a temporally extended pattern such as a bird song or a sentence. Yet the timing of neuronal events seems to be absolutely crucial to just about everything a nervous system does. We do not know how movements can be sequenced or how sequences can be modified or how movement decisions are made. We do not know the extent to which mammalian nervous systems are input-output devices and the extent to which their activity is intrinsic in a way utterly different from any known computer. We do not know why we sleep and dream, what sensory images are, and how the appropriate ones come into being when needed.

Despite confident announcements in evolutionary psychology about genes "for" this and that mental capacity, in fact we do not know what or how much of the neuronal structure in mammals is genetically specified, although simple arithmetic tells us that it cannot be specified synapse by synapse. Specialization of structure does of course exist in nervous systems, but the nature and degree of top-down influence in perception remains unclear, as does cross-modal integration and cognitive coherence. Whether nervous systems have *modules*, in the sense that some psychologists define the word, looks highly unlikely, although how to characterize an area of specialization, if it is not a "module" in the classical sense, is puzzling. The puzzle will not be resolved merely by cobbling together some precise definition. To serve a scientific purpose, the definition has to grow out of the empirical facts of the matter, and many relevant facts we just do not have our hands on yet. Virtually every example of modularity (in the classical sense) that has ever been explored turns out, when experiments alter the developmental conditions, not to be a modular entity except in the most attenuated of senses. And, to continue, we do not know what functions are performed by the massive numbers of back projections known to exist just about everywhere.

This is a lot not to know. And it highlights the need for an explanatory exoskeleton adequate to these issues. Partly, because the profound, exoskeletal questions remain, neuroscience is both tremendously exciting and also faintly exasperating. Sometimes one is moved to wonder whether we are asking the neural equivalent of what makes the crystal spheres turn or what the weight of phlogiston is or why God decided to make disease a punishment for sin. In his opening remarks at the 1999 Centennial Fellowship Symposium, Endel Tulving warned that our common sense—our intuitive hunches—about what the brain is doing and how it is doing it are, in many cases, going to be flat-out wrong. I share Tulving's prediction, and it is tantalizing to wonder just how hypotheses and theories in cognitive neuroscience will look 100 years from now.

My handful of neuroscience questions is unsystematically presented and insufficiently developed, but my intent here is merely to convey a simple point: There is a great deal of foundational science that is still to be done on nervous systems. By emphasizing *questions* as opposed to *answers*, I emphatically do not mean to diminish the brilliant progress in neuroscience. Certainly, much about the basic components—neurons—has been discovered; and without these discoveries, functional questions about how we see, plan, decide, and move could not fruitfully proceed. Much about the anatomy and the physiology at the systems level has also been discovered. That too is absolutely essential groundwork. I have no wish to chastise neuroscience. Rather, I want affectionately to view it at arm's length and to cheer on those who are willing to tackle some of the exoskeletal questions. Because there is much of a fundamental nature that we do not know about the brain, there is a lot of frontier territory. There is still room—and need—for pioneers.

5

Taking a Snapshot of the Early Universe

John E. Carlstrom

INTRODUCTION

Starting as children, we are driven to explore our environment and constantly expand our horizons. The field of cosmology coordinates this exploration, allowing each generation to start its exploration at the frontier of the previous generation. Cosmology strikes a fundamental chord in all of us; we are struck with the enormity of the universe and are humbled by our place within it. At the same time we are empowered with our ability to obtain answers to fundamental questions of the universe. Peering back to the early universe and taking a snapshot does more than let us measure cosmological parameters—it enriches us and satisfies a fundamental desire in all of us.

Copernicus at the beginning of the Renaissance set in motion a fundamental change in our understanding of our place in the universe by showing that the Earth circles the Sun. Today, cosmologists are also setting in motion a fundamental change by showing that the matter we know about, the stuff that makes up the Sun, Earth, and even ourselves, only accounts for a small fraction, perhaps as little as 10 percent, of the mass of the universe.

Cosmology today is at a crossroads where theory and experiment are coming together. Astrophysicists are narrowing in on the Hubble constant, the rate of expansion of the universe. Theorists have developed a

Department of Astronomy and Astrophysics, University of Chicago

standard cosmological model for the origin of structure in the universe. They have also shown that the key to understanding the universe is contained in the cosmic microwave background radiation, the relic radiation of the Big Bang, motivating experimentalists to build ever-more sensitive experiments. Through studies of the microwave background, theorists and experimentalists are on the cusp of testing rigorously the standard model and determining precise values of cosmological parameters, which will soon tell us the age of the universe, whether it will expand forever or close in upon itself, and how much mass is in a form still unknown to physicists.

The cosmic microwave background photons (particles of light) travel to us from across the universe, from a time when the universe was in its infancy, less than a ten-thousandth of its present age. Small differences, of an order of a few hundred-thousandths, in the intensity of the microwave background carry information about the structure of the infant universe. The properties of the cosmic microwave background also make it an incredible backlight with which to explore the universe; it is strong, even a radio can pick it up, and it is extraordinarily isotropic. If our eyes were sensitive to microwave radiation, we would see a uniform glow in all directions. However, just as the stained glass of a cathedral window creates a beautiful display of color and shape from a common backlight, large-scale structures in the universe alter the intensity and spectrum of the microwave background. The results can be stunning, and, moreover, much can be learned about the universe from detailed studies of the display. These microwave eyes, however, would have to be very sensitive to see it, as the display has very little contrast, with only a tenth of a percent or less deviation from the uniform glow of the background.

My collaborators and I built a sensitive system to image the display created by the passage of the cosmic microwave background radiation through the gas contained within clusters of galaxies. The effect is known as the Sunyaev–Zel'dovich effect after the two Russian astrophysicists who predicted it in 1972. Using the powerful technique of radio interferometry, our system has allowed us to make, for the first time, detailed images of the effect. With these images we are able to measure distances to clusters and therefore obtain a completely independent measure of the Hubble constant, using a technique based solely on the physics of the cluster gas. Our Sunyaev–Zel'dovich effect data also allow us to estimate the ratio of ordinary matter to the total mass of a galaxy cluster. Combined with results from primordial nucleosynthesis calculations and elemental abundance measurements, our results allow an estimate of the mass density of the universe. The mass density that we find is close to the critical value that marks the transition between a universe which expands forever and one which eventually will collapse upon itself due to the gravitational attraction of the matter.

So far we have limited our Sunyaev–Zel'dovich effect observations toward known clusters of galaxies. However, we wish to exploit a feature of the effect that is immensely powerful: The strength of the observed effect is independent of the distance to the cluster. The effect is therefore a unique probe of the distant universe and, in principle, can be used to place firm observational constraints on the evolution of structure in the universe. Currently, however, it takes several days to obtain a clear image of the effect toward a rich cluster of galaxies. By improving our instrument, we plan to decrease this time by more than a factor of 10 and begin a thorough exploration of the distant universe.

Based on the success of our Sunyaev–Zel'dovich effect imaging system, my collaborators and I have begun a much more ambitious experiment. We plan to take a detailed snapshot of the early universe. I am working with collaborators at the University of Chicago and at Caltech to build a pair of dedicated interferometers to image the small intensity differences, of an order of a few hundred-thousandths, in the cosmic microwave background radiation itself. Because this small anisotropy in the background radiation provides the key to understanding the origin and evolution of structure in the universe, these new instruments will open a new window on the universe. Although the instruments are designed to provide answers to some of the most pressing questions in modern cosmology, we recognize that discoveries in astrophysics are driven by new techniques and instrumentation. We fully expect that the new window on the universe opened by these instruments will provide fascinating and unexpected views.

THE EARLY UNIVERSE

Winding the Clock Backward

The early universe was much simpler than it is today. Yet from these relatively simple beginnings arose the vast multitude of complex phenomena that are present today.

Using ever-more powerful telescopes, astronomers have explored deep into the universe, looking far back in time. They have shown that the universe is expanding, with galaxies flying apart from each other with a speed proportional to their separation. The Hubble constant is the ratio of the recessional speed of a galaxy to its distance from our galaxy. Experiments have shown that the Hubble constant is a global constant; all galaxies share the same linear relationship between their recessional velocity and distance. The exact value of the Hubble constant is difficult to determine. Precise recessional velocities are measured easily using spectroscopically determined Doppler shifts, but astronomical distances are problematic, especially for distant objects. Nevertheless, the Hubble constant is now known to lie roughly in the range 50–85 km s^{-1} Mpc^{-1}. Knowing the expansion rate, astronomers can predict the time when the expansion started. This time is simply the inverse of the Hubble constant. For 65 km s^{-1} Mpc^{-1}, one over the Hubble constant equals 15 billion years. Accounting for the gradual slowing of the expansion due to the gravitational attraction of the matter gives an approximate age for the universe of 10–15 billion years.

What would we see if we were able to view the universe at earlier and earlier times? As we go back in time we notice that the universe is denser and hotter. We pass through the epoch when galaxies were formed, beyond the formation of the first stars, to a time when electrons and protons were combining to form hydrogen. The universe is about 300,000 years old, a tiny fraction of its present age. The cosmic microwave background radiation was emitted during this epoch. At earlier times the universe was made up of a simple fluid of photons (massless particles of light), subatomic particles, and the mysterious dark matter, matter which astronomers know about only by its gravitational pull on ordinary matter.

Properties of the Fluid

The dynamics of the fluid, or plasma, of the early universe were driven primarily by two forces: the gravitational attraction of the massive particles (baryons) and internal pressure caused by the continual scattering of photons by free electrons. The scattering keeps the photons coupled tightly to the plasma. Density perturbations will initiate a gravity-driven collapse. As the density increases during the collapse, the rate of photon–electron interactions—and therefore the pressure resisting the collapse—increases. This situation leads naturally to oscillations, actually acoustic (sound) oscillations.

Acoustic oscillations will be excited for regions spanning a broad range of sizes. The larger the size, however, the longer they take to develop. Thus there is a natural limit to the largest region that will begin to collapse, roughly the age of the universe times the speed of light. The cooling of the universe as it expands caused the electrons and protons to form hydrodgen when the universe was only 300,000 years old. This period, referred to as the epoch of "recombination," even though it is in fact the first "combination" of electrons and protons, significantly affected the evolution of structure in the universe. With the electrons confined within atoms, the photons no longer interact strongly with matter and begin streaming freely through the now transparent universe. The decoupling of the photons and matter causes the fluid to lose its internal support against gravity. Structures are now able to collapse further, leading eventually to the rich structure of the present universe.

THE COSMIC MICROWAVE BACKGROUND

The photons, which have been streaming freely through the universe since the epoch of recombination, also known as the period of last scattering, are detected today as the cosmic microwave background. Imprinted on the background as small fluctuations in its intensity, of an order a few hundred-thousandths, are the signatures of the acoustic oscillations present at the period of recombination. An image of the anisotropy in the microwave background provides a snapshot of the early universe and contains a wealth of information.

Astronomers infer the properties of astrophysical phenomenon by studying the property of the light received by their telescopes, its intensity, spectrum, and distribution on the sky. Astronomers also use scattered light to learn about the scattering medium. When the light has been scattered repeatedly, its direction, and eventually its energy, no longer carry information of the original source. Therefore, studying the cosmic microwave background provides information on the universe at the time of last scattering. It also provides information about the objects the microwave background photons encountered as they traversed the universe.

The spectrum of the cosmic microwave background can tell us a great deal about the early universe. The continual interaction of photons and electrons and other particles led to a condition of thermal equilibrium in the early universe. In equilibrium the distribution of the energies of the particles (i.e., velocities) and of the photons (i.e., wavelengths) are well known and described fully by a Planck spectrum with only knowledge of the temperature (see Figure 1). We are thus able to predict the spectrum of the background at the epoch of recombination; its spectrum is predicted to be a Planck spectrum for a temperature of 3,000 K.

Because of the expansion of the universe, the spectrum of the background radiation is different today than at the epoch of recombination. As space expands, so do the wavelengths of the photons traversing the universe. If the original spectrum is described by a Planck spectrum, then the resulting spectrum would still be described fully by a Planck spectrum, but at a lower temperature. The Cosmic Background Explorer (COBE), a NASA satellite designed to study the cosmic background radiation, measured its spectrum and found that it is indeed described extremely well by a Planck spectrum for roughly 3 K. This spectacular re-

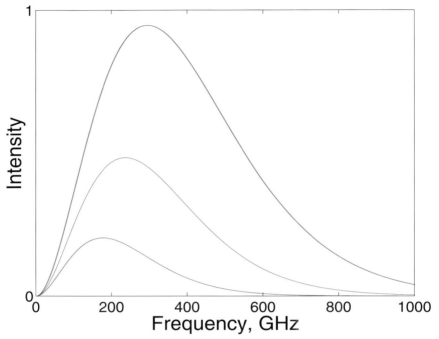

FIGURE 1 Planck spectrum. The spectrum of radiation in thermal equilibrium is determined only by its temperature. It is independent of the composition of the object. A spectrum for a given temperature encloses all spectra with lower temperatures. This fact allows a temperature determination with a measurement at only one wavelength. The cosmic microwave background radiation is well described by a Planck spectrum for a temperature of roughly 3 degrees above absolute zero indicated by the red curve.

sult illustrates the ability to use the cosmic microwave background radiation to directly measure the properties of the early universe.

Ripples Across the Sky

The horizon—the distance spanned traveling at the speed of light for the age of the universe—at the epoch of recombination, when the cosmic microwave background photons last scattered, subtends an angle of roughly 1 degree today. Thus temperature fluctuations in the microwave background on angular scales smaller than a degree carry information about the earliest evolution of structure in the universe. On angular scales larger than a degree, the temperature fluctuations carry information on the density inhomogeneities for regions that were disconnected—so far apart that even light traveling for the age of the universe could not carry information from one region to another. Such large-angular-scale temperature fluctuations in the cosmic microwave background can reveal the primordial density inhomogeneities in the universe.

Thus the measurement of temperature anisotropies at 7 degrees and greater angular scales by the COBE satellite caused great excitement as it provided a look, for the first time, at the primordial density inhomogeneities of the early universe. Since the COBE results were announced in 1992, many experimental groups have announced detection of temperature anisotropies, both on large angular scales and on degree and smaller angular scales. A compilation of these results is shown in Figure 2, where the vertical axis is the average level of the temperature fluctuations, and the horizontal axis is the angular scale over which the temperature fluctuations were measured. The horizontal axis is labeled in terms of angular multipole moments, which is just a convenient method for ordering the angular separations (an angular separation of θ is given roughly by $100°/l$. Such a plot is referred to as an angular power spectrum. Note that the measurements show an apparent increase in the level of the background fluctuations at angular scales smaller than 1 degree, $(l \geq 100)$ which corresponds to the size of the horizon at the surface of last scattering.

On the theoretical side, the expected temperature fluctuations in the cosmic microwave background have been computed for a wide range of cosmological models. These models, of course, are incapable of predicting whether a particular direction on the sky is hot or cold, but instead they predict the expected magnitude of the temperature differences, on average, as a function of the angle of separation between two positions on the sky. The quantitative results are best shown with the aid of the angular power spectrum (see Figure 2). The predicted power spectrum shows a series of peaks and troughs, which are now understood as signatures of the acoustic oscillations in the photon pressure-supported fluid of the early universe. The first peak in the plot at roughly a half a degree $(l\sim200)$ corresponds to temperature fluctuations caused by the largest regions to reach maximal compression at the time of decoupling. The temperature in these overdense regions is elevated slightly. The formation of these regions must also create underdense regions, for which the temperature is depressed slightly. The net results are ripples in the microwave sky. The power spectrum provides a way to present the magnitude of the ripples as a function of the subtended angle between neighboring hot (dense) and cold (rarefied) regions.

The next peak in the power spectrum is due to smaller regions that had just enough time to reach maximal rarefication (i.e., advanced by a half-cycle). These regions are slightly cooler, but still lead to a peak in the background power spectrum because it is a measure of the fluctuations in the temperature. The process continues with each successive peak corresponding to oscillations advanced by a half-cycle to those of the previous peak. The "surface" of last scattering actually has a finite thickness, and therefore higher-order peaks in the power spectrum (which correspond to much smaller regions) appear much weaker because they represent an average of both hot and cold regions lying along the line of sight.

With this simplified theoretical picture in mind, one can see that sev-

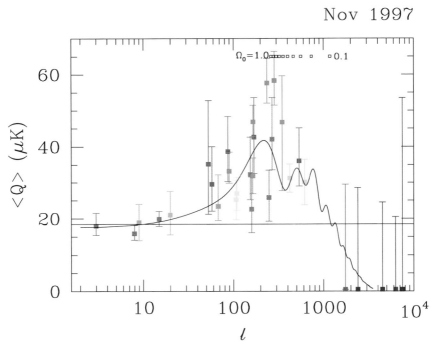

FIGURE 2   The experimental state of cosmic microwave background anisotropy measurements as of November 1997. The experimental results are shown as colored squares and error bars. The black curve showing a peak near $l = 200$ is the theoretical prediction for the standard inflationary model. A horizontal line drawn at the level of the COBE results at l less than 20 clearly underestimates the level of detection at degree angular scales ($l\sim200$); there is strong evidence that excess signal is detected at these angular scales. The small square boxes near the top of the figure indicate the location of the first peak predicted by cosmological models for differing values of the mass density of the universe. Thus far no experiment has had the necessary sensitivity and resolution in l to distinguish between models and to extract the cosmological parameters. Figure courtesy of M. White.

eral cosmological parameters can be extracted directly from the cosmic microwave background power spectrum. For example, the amplitude of each peak is related to the strength of the corresponding oscillation, which is determined by the density of ordinary matter. The physical scale associated with the first peak of the power spectrum is the horizon size at the time of decoupling. The observed angular size of this feature depends on the curvature of the universe, which depends on the sum of the total mass density of the universe and Einstein's cosmological constant. If the universe is flat, as predicted by the popular inflationary model for the early universe, then the first peak will be located at roughly a half a degree (l~200). For an open universe, one that will expand forever, space is curved and the first peak shifts to smaller angular scales.

It is remarkable that by making sensitive measurements of the tiny temperature fluctuations in the cosmic microwave background, we will learn the ultimate fate of the universe, whether it will expand forever, collapse upon itself, or whether the universe is precisely balanced between these fates.

The Sunyaev–Zel'dovich Effect: Mass and Age of the Universe

As the cosmic microwave background photons travel from the surface of last scattering to the observer, they may suffer interactions with intervening ionized matter. Just as the stained glass in a cathedral window creates a beautiful display of color and shape from a common backlight, these interactions alter the intensity and spectrum of the microwave background. Of particular interest is the effect that occurs when the background photons travel through a cluster of galaxies.

Galaxy clusters are enormous objects, containing as much as a thousand trillion times the mass of our Sun with the vast majority of the mass lying outside of the galaxies themselves in the form of intergalactic hot plasma and other as yet unknown forms of matter. This hot (of the order of 100 million Kelvin) x-ray-emitting plasma is spread uniformly throughout the cluster. A cosmic background photon traveling at the speed of light takes several million years to cross a cluster. Even so, it has only about a 1 percent chance of suffering an interaction within the plasma.

The scattering of the microwave photons by electrons in this hot plasma results in a small, localized distortion of the microwave background spectrum at the level of a thousandth or less. This spectral distortion of the microwave background is known as the Sunyaev–Zel'dovich effect. Although the effect is rather weak, it is much stronger than the anisotropy imprinted on the microwave background from the surface of last scattering. This and the fact that it is a localized effect make it possible to separate the two phenomena.

Because they are among the largest objects in the universe, galaxy clusters provide powerful probes of the size, structure, and origin of the universe. Their cosmological significance is augmented by the fact that an independent determination of their distances, and therefore of the Hubble constant, can be made if accurate measurements of both the Sunyaev–Zel'dovich effect and the underlying x-ray emission can be made. In addition, x-ray and Sunyaev–Zel'dovich data can be used to measure the mass density of the universe by determining the ratio of ordinary to total mass in galaxy clusters and by determining the properties and number density of distant clusters.

The expected spectral distortion of the cosmic microwave background when observed toward a galaxy cluster is shown in Figure 3, relative to the undistorted spectrum. At radio frequencies, the Sunyaev–Zel'dovich effect produces a small decrement in the observed intensity. Amazingly, the Sunyaev–Zel'dovich effect predicts that an image of a rich cluster of

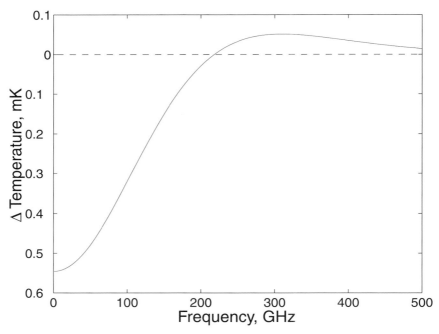

FIGURE 3 The Sunyaev–Zel'dovich effect spectral distortion of the cosmic microwave background spectrum due to the hot ionized gas associated with a massive cluster of galaxies (red line).

galaxies (an object with more than a trillion times the luminosity of the Sun) made at radio wavelengths should show a hole in the sky relative to the smooth glow of the microwave background.

The key to understanding the determination of the Hubble constant, and therefore the age of the universe, from a combination of the Sunyaev–Zel'dovich effect and x-ray data lies in their different dependencies on the density of electrons in the hot gas. Basically, the magnitude of the Sunyaev–Zel'dovich effect is proportional to the total number of electrons along the line of sight independent of how dense the gas may be. The magnitude of x-ray emission, however, is strongly dependent on the density of the gas. If the same amount of gas was compressed to a higher density, then its x-ray emission would become much stronger, whereas the Sunyaev–Zel'dovich effect would be unaffected. Combining both x-ray and the Sunyaev–Zel'dovich effect measurements of the gas thus allows a determination of its density structure, which in turn allows a determination of the size scale of the gas region. A comparison of this size with the observed angular extent of the cluster gas gives a measure of the distance to the cluster. The Hubble constant is given directly by combining the distance with the observed recessional velocity of the cluster.

A measurement of the Sunyaev–Zel'dovich effect also provides a direct measure of the mass of gas (ordinary matter) if the gas temperature, determined by x-ray spectroscopy, is known. Assuming the gas is roughly in equilibrium, not collapsing or expanding, then given its temperature, it is straightforward to predict the *total* mass, including the dark matter, that is required to bind the gas to the cluster through gravitational forces. Because clusters formed from large volumes, it is reasonable to expect that the ratio of their mass of ordinary matter to their total mass reflects the overall ratio for the universe. Cosmologists have narrowed the acceptable range for the mass fraction of ordinary matter in the universe. A mass fraction measured using the Sunyaev–Zel'dovich effect thus leads to an estimate of the total mass density of the universe.

A remarkable property of the Sunyaev–Zel'dovich effect is that the observed decrement is independent of the distance to the cluster. This is understood by recognizing that the ratio of the Sunyaev–Zeldovich effect temperature change to the temperature of the undistorted microwave background is a property of the cluster. Even though the temperature of the background was higher when the microwave photons interacted with the cluster gas (recall that the temperature of the microwave background was 3,000 K when it left the surface of last scattering and it is only 3 K now), both the temperature decrement due to the Sunyaev–Zel'dovich effect and the temperature of the microwave background decrease with the expansion of the universe in a way that maintains their ratio. Thus the magnitude of the Sunyaev–Zel'dovich effect is independent of the distance to the cluster. This property is in sharp contrast to the observed emission from an object, such as x-ray emission from the gas or optical emission from the galaxies, for which the observed flux rapidly fades with distance.

Large-scale surveys of the Sunyaev–Zel'dovich effect, therefore, offer the potential to probe the distant universe and to take an inventory of all massive clusters. The data from such a survey would greatly aid in our understanding of the formation of large-scale structure in the universe, allowing severe tests of models of structure formation and evolution because predictions for the number, masses, and morphology of distant clusters differ widely between competing models. For example, the current factor of 5 in the uncertainty of the mass density of the universe leads to an order of magnitude difference in the predicted number of distant clusters.

Since Sunyaev and Zel'dovich made the first theoretical prediction of a cluster-induced distortion in the cosmic microwave background in 1972, numerous groups have struggled to detect this effect. Until quite recently, the effect has been detected only for a small number of clusters. That situation changed dramatically with the introduction of interferometers specifically tailored for accurate imaging of the Sunyaev–Zel'dovich effect in galaxy clusters. As I show below, my group has made substantial contributions in this area.

EXPERIMENTAL CHALLENGES

The cosmic microwave background radiation is "detected" easily, although it is difficult to distinguish from other sources of noise. The snow on a television set, or the hiss of a radio, when tuned away from a station, has a significant contribution from the microwave background. In 1964 Arno Penzias and Robert Wilson were investigating the source of radio interference and discovered that the excess noise was coming from the heavens and appeared isotropically distributed. They had detected the cosmic microwave background radiation, the relic radiation of the Big Bang, for which they were awarded the Nobel Prize.

What is difficult to detect, however, are temperature differences in the microwave background—differences in the intensity of the radiation—between one part of the sky and another. The largest temperature anisotropy of the background is due to the motion of our solar system with respect to the inertial frame of reference of the background. Our solar system is moving with a speed of 370 km s^{-1} with respect to the background. This fact is inferred from a positive change in the temperature of the microwave background at a level of one-tenth of a percent when looking along the direction of our motion and a negative change of the same magnitude when looking in the opposite direction. The COBE satellite has shown that, after accounting for this contribution, the departures from a perfectly smooth background on angular separations as close as 7 degrees are only of an order of several millionths.

There are a number of challenges facing experimentalists now and in the near future. First and foremost is a measurement of the anisotropy spectrum of the cosmic microwave background at angular scales from a few arcminutes to a degree ($100 < l < 2000$), with sufficient sensitivity to distinguish between cosmological models and to extract the cosmological parameters. Images of the microwave background are needed with sufficiently high quality to test not only model predictions of its distribution, but also to allow features to be inspected and compared with observations using a variety of techniques and wavelengths. At arcminute angular scales ($l{\sim}3000$), images of large regions of the sky to inventory all galaxy clusters are needed to allow severe tests of models of structure formation and evolution.

The large scientific rewards for meeting these challenges have motivated many excellent research teams. There are no fewer than 12 new experiments planned, and both NASA and the European Space Agency have selected cosmic microwave background satellite missions to scan the entire sky. Building on the success of our Sunyaev–Zel'dovich effect imaging program, my team plans to tackle these challenges using novel interferometric arrays. As I discuss below, the inherent stability of interferometry will allow us to make high-quality images of a large fraction of the sky from superb ground-based sites.

Interferometry: A Different Way of Seeing

Radio interferometry was developed to improve the angular resolution of radiometric observations by taking advantage of the wave-like nature of light. The technique also provides a number of benefits for making highly sensitive images at any resolution. In particular, it shows great promise for imaging the microwave background. However, new instruments must first be built because traditional interferometers, designed for high resolution, are insensitive to the angular scales needed to image the cosmic microwave background. Before discussing these new instruments, it is instructive to consider how traditional interferometers work.

The angular resolution of a telescope is limited by diffraction because of the wave-like properties of light. Imagine using a buoy placed in a lake to detect waves in the water. The buoy bobs up and down clearly detecting water waves, but what information does it provide regarding the direction of the waves? To help determine the direction, we might place several buoys next to each other in a line. In this case, if all the buoys rise and fall synchronously, we would know that the wave was propagating from a direction perpendicular to the line of buoys. How well can we determine the direction to the source of the waves, or equivalently, how well can we determine the orientation of the wave front? It must depend on two quantities: the length L of the buoy array and the wavelength λ of the wave being detected. Roughly, we can expect to determine the direction with an uncertainty of the order of λ/L. If we want to determine the direction of a wave with a particular λ to high accuracy, we must increase L accordingly.

Consider a radio telescope. They can be very large, with diameters up to 100 m; however, they also operate at long wavelengths. Most radio telescopes were designed for wavelengths ranging from 1 m to 1 cm, although there are now extremely precise "radio" telescopes designed for wavelengths as short as 1 mm and a few for wavelengths as short as 0.3 mm. Consider the size of a radio telescope operating at 1 m that would be capable of the same resolution (i.e., same λ/L) as an eye looking at blue light. The pupil is only 4 mm in diameter, whereas blue light has a wavelength of ~400 nm, yielding $\lambda/L = 0.0001$, which corresponds to better than 1 arcminute, or the angle subtended by a quarter placed 100 m away. To achieve the same resolution at a wavelength of 1 m requires a radio telescope with a diameter of 8 km! Clearly, this is not practical.

The buoy example offers another possibility. If there are gaps in the line of buoys, we can still determine the direction of the wave. The analogous solution for radio telescopes is to use arrays of radio telescopes. This technique, called radio interferometry, is used routinely in radio astronomy. Arrays such as the Very Large Array (VLA) operated by the U.S. National Radio Astronomy Observatory are designed exclusively for interferometry. The 25-m-diameter telescopes of the VLA can be separated by more than 50 km, giving an angular resolution of 4 arcseconds at an operating wavelength of 1 meter. At its shortest operating wavelength of 7 mm, the VLA provides an angular resolution of three-hundredths of an arcsecond, about three times better than the Hubble Space Telescope! As impressive as this may seem, this resolution is crude compared with that achieved using very long baseline interferometry (VLBI). I have been part of a team of astronomers that has used a network of radio telescopes, located across several continents, to achieve an angular resolution of five ten-thousandths of an arcsecond.

If arrays work so well, why build large telescopes at all? The reason is simple: to collect as much light as possible. The amount of power received per square meter (referred to as flux) from a source that subtends an angle of only five ten-thousandths of an arcsecond is tiny, unless the source is extremely bright. In fact, only the brightest and most compact objects can be seen using VLBI. A crude estimate of the minimum brightness object detectable at this resolution is about 10 million K, much brighter than the Sun. VLBI, however, is ideally suited for observations of the energy released from matter as it is accreted by massive black holes found in the center of active galaxies. We have observed objects as bright as a trillion degrees Kelvin.

Operating at shorter wavelengths improves the brightness sensitivity of an interferometer. Why this is the case is illustrated by the Planck curves in Figure 1; for a given temperature an object simply emits more energy at shorter wavelengths. The current state-of-the-art of traditional radio interferometry is operation at submillimeter (submm) wavelengths. My collaborators and I built the first submm-wave interferometer by linking the 15-m James Clark Maxwell Telescope (JCMT) and the 10.4-m Caltech Submillimeter Observatory (CSO) located at 14,000 ft near the summit of Mauna Kea, Hawaii. Our goal was to determine the nature of the excess submm-wave emission associated with young stars. Specifically, we wanted to know if the emission was from dusty circumstellar disks, similar to the early solar system. These young systems are hidden from view at optical wavelengths by the interstellar clouds in which they form.

An interferometer is needed to measure these protoplanetary disks. At the distance of the nearest stellar nurseries, the Taurus and Ophiucus clouds, the solar system would subtend an angle of an arcsecond or less, much smaller than the 10-arcsecond resolution of the JCMT telescope, the largest submm-wave telescope in the world. However, using the half-arcsecond resolution enabled by the 165-m separation of the CSO and JCMT telescopes, we have measured dimensions and masses of circumstellar disks for several stars. We find their properties quite similar to those inferred for the early solar system.

Benefits of Interferometry

Interferometry offers a number of attractive features even when high angular resolution is not desired. Namely, it offers high stability and sensitivity and the ability to produce two-dimensional images directly. These benefits arise naturally due to interferometers taking full advantage of the wavelike nature of light, using both amplitude and phase information. As shown in Figure 4, each pair of telescopes in an interferometer is sensitive to a particular "ripple" on the sky. The cosmic microwave background power spectrum (Figure 2) is a decomposition of the microwave sky into waves, with the wavelengths specified by l. This decomposition of a picture into waves is graphically illustrated in Figures 5–7. Although an interferometer measures directly the power spectrum, it also measures

Interferometers detect ripples on the sky

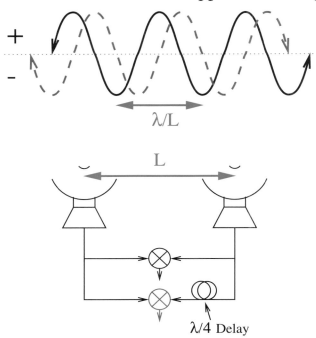

FIGURE 4   The response of a two-element interferometer is a sinusoid pattern on the sky. Signals received by the telescopes are transmitted to a central location where they are multiplied together. The product depends on the difference in path lengths from the source to the telescopes and on to the multiplier. The output of the multiplier therefore depends on the position of the source in the sky. The solid blue curve illustrates the dependence of the detected signal on the position of the source in the sky. Note that it is a sinusoid with an angular spacing set by the wavelength of the light divided by the separation of the telescopes. Thus ripples on the sky are measured directly. One may worry about not detecting a source where the pattern is zero. For this reason, a second multiplication is performed with an additional quarter-wavelength of path inserted into one of the signal paths. The detected signal from this multiplication is sensitive to the same ripple on the sky but is offset by a quarter-cycle as shown by the dashed red line. The ripple pattern causes the interferometer to difference simultaneously one part of the sky against another, allowing extremely sensitive observations. The pattern is also ideal for measuring ripples in the cosmic microwave background, or what cosmologists call the angular power spectrum of the microwave background. The many pairs of telescopes in an array, such as the Degree Angular Scale Interferometer, each measure a particular ripple pattern. Outputs from the two multipliers fully determine both the *amplitude and phase* of the ripple. Once the amplitudes and phases are measured, the ripple patterns, like a hologram, can be used to generate a picture of the sky (also see Figures 5–7).

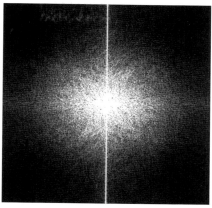

FIGURE 5 **Left panel**: A picture of colleague, José, shown in the representation that we are all used to seeing. **Right panel**: An alternative representation of the picture of José where the intensity corresponds to the magnitude of different "ripples" in the picture of José. Position in this representation corresponds to a particular wavelength and direction for a ripple, or wave. This representation is referred to as the Fourier transform of the picture, named after the mathematician who showed that no information is lost by a decomposition into waves. Each wave has an associated phase (not shown) that indicates where its peaks and troughs should be lined up when the waves are added together to reconstruct the original picture. An interferometer measures directly the amplitude and phase of these ripples.

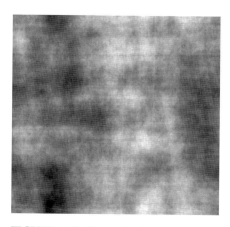

FIGURE 6 **Left panel**: A reconstruction of the picture of José from the amplitude of the waves in the right panel of Figure 5, but using random phases to align the waves. **Right panel**: A reconstruction using the correct phase information. Note that *both panels have the same power spectrum*. The figures clearly show that the power spectrum does not contain all of the information. The phase of a wave, which is measured directly by an interferometer, is just as important as its amplitude.

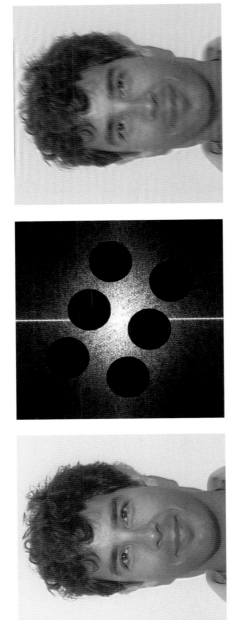

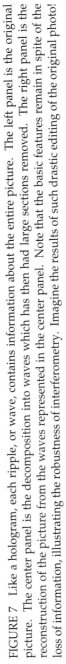

FIGURE 7 Like a hologram, each ripple, or wave, contains information about the entire picture. The left panel is the original picture. The center panel is the decomposition into waves which has then had large sections removed. The right panel is the reconstruction of the picture from the waves represented in the center panel. Note that the basic features remain in spite of the loss of information, illustrating the robustness of interferometry. Imagine the results of such drastic editing of the original photo!

the phase. As illustrated in Figure 6, the power spectrum alone is not sufficient to reconstruct a picture.

The positive and negative series of troughs and peaks of the response of an interferometer lead to exceptional stability and low systematics. A well-designed interferometer can take data for essentially an unlimited time without being limited by systematics; we have observed a single cluster for 14 days to image the Sunyaev–Zel'dovich effect.

INTERFEROMETERS AND COSMIC MICROWAVE BACKGROUND

Imaging the Sunyaev–Zel'dovich Effect

Galaxy clusters are very large containing over a 100,000 times the volume of our galaxy. Even so, a cosmic background photon passing through a cluster has only about a 1 percent chance of suffering an interaction with an electron. The resulting Sunyaev–Zel'dovich effect that we wish to observe is therefore quite weak and extends smoothly over arcminute angular scales. Using standard radio telescopes, it has proved difficult to detect, let alone map, the Sunyaev–Zel'dovich effect.

My collaborators and I decided that the best way to detect and map the Sunyaev–Zel'dovich effect would be to use radio interferometry. It is straightforward to write down the specifications for such an interferometer. Its operating wavelength should be chosen to maximize the Sunyaev–Zel'dovich effect signal (see Figure 3), minimize noise from our atmosphere which increases at short wavelengths, and to minimize emission from galaxies which increases at long wavelengths. A wavelength of about 1 cm is ideal. Once the wavelength is decided, the minimum separation of the telescopes is chosen so that the angle subtended by the cluster (i.e., 1 arcminute) fits well within λ/L (see Figure 4). The telescope diameters must, of course, be less than L. For 1 arcminute and λ equal to 1 cm, a good choice for L is about 6–10 m. So we want an array of roughly 8-m telescopes outfitted with high-quality 1-cm receivers. It should not come as a surprise, because interferometers have traditionally been built for high (arcsecond) resolution, that such an array did not exist.

Arrays with 6-m- and 10-m-diameter telescopes *do* exist, however. They are used for making arcsecond resolution observations at millimeter wavelengths. By outfitting these millimeter wave arrays with centimeter wave receivers, we created a nearly ideal instrument for detecting and making detailed pictures of the Sunyaev–Zel'dovich effect. We built state-of-the-art low-noise receivers that we mount for roughly a month at a time during the summer on the six 10.4-m telescopes of the Owens Valley

FIGURE 8   The BIMA millimeter-wave array with the 6.1-m diameter telescopes positioned in a compact configuration for observations of the Sunyaev–Zel'dovich effect.

Radio Observatory (OVRO) millimeter array operated by Caltech and on the nine 6.1-m telescopes of the Berkeley Illinois Maryland Association (BIMA) millimeter array. We move the telescopes as close as possible when observing the Sunyaev–Zel'dovich effect to obtain the best match between the resolution of the interferometer (λ/L; see Figure 4) and the roughly 1-arcminute angle subtended by a distant galaxy cluster. A photograph of the BIMA array taken while we were observing the Sunyaev–Zel'dovich effect is shown in Figure 8. The telescopes are close enough to collide! Luckily, safeguards in the control software and collision sensors installed on the telescopes have prevented such a disaster, although we have had several scares.

Our interferometric Sunyaev–Zel'dovich effect system works beautifully. We have gone beyond the point of just trying to detect the Sunyaev–Zel'dovich effect to producing high-quality images. In Figure 9 we show our image of the Sunyaev–Zel'dovich effect overlaid on the x-ray emission for the cluster CL 0016+16. In Figure 10 we demonstrate the independence of the Sunyaev–Zel'dovich effect on the distance to the cluster. Shown are three clusters with distances spanning nearly a factor of 5. The observed Sunyaev–Zel'dovich effects are about the same, whereas the

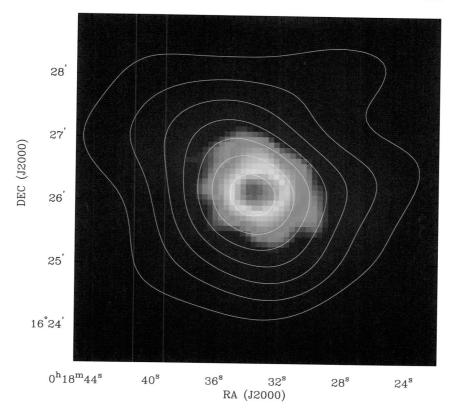

FIGURE 9 An image of the Sunyaev–Zel'dovich effect (contours) and x-ray emission (false color) due to the hot gas associated with the galaxy cluster CL 0016+16. The radio observations of the Sunyaev–Zel'dovich effect were obtained using the BIMA and OVRO interferometric arrays. The x-ray observations were obtained with the ROSAT satellite. The Sunyaev–Zel'dovich effect is contoured in steps of negative intensity. Because of the Sunyaev–Zel'dovich effect, the cluster, which has a luminosity roughly a trillion times the Sun, appears as a hole in the microwave sky.

observed x-ray flux of the most distant cluster is roughly a 100 times weaker than the flux of the closest cluster.

With quality data for more than 20 clusters, we are now in the enviable position of using the Sunyaev–Zel'dovich effect to estimate the Hubble constant and the mass density of the universe. Our preliminary analyses give a value of about 65 km s^{-1} for the Hubble constant.

For objects as massive as galaxy clusters, it is reasonable to expect that

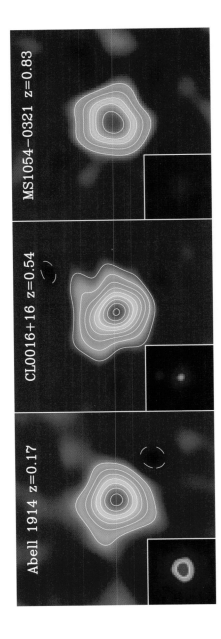

FIGURE 10   Images of the Sunyaev–Zel'dovich effect toward clusters at different distances.   The images are shown in false color with contours overlaid—the same contour levels are used for each image.   Red indicates lower temperatures.   The redshift z of the cluster is proportional to its distance.   The corresponding x-ray emission observed toward each cluster is shown in the inset of each panel with the same false-color x-ray intensity scale used for each cluster.   In contrast to the Sunyaev–Zel'dovich effect, the x-ray flux from these clusters exhibits a strong decrease with redshift (distance).

the ratio of ordinary matter to the mysterious dark matter would be a fair representation of the value for the entire universe. Using our Sunyaev–Zel'dovich effect data for several clusters gives a ratio of 12 percent, where we have assumed the Hubble constant is 65. Amazingly, nearly 90 percent of the mass in the universe is in a form that physicists still do not understand!

Cosmologists do, however, know the mass density of ordinary matter in the universe. Taking our value of 12 percent for the mass fraction of ordinary to total matter and the value for the mass density of ordinary matter, we have solved for the total mass density of the universe. Our preliminary results suggest that the total mass density of the universe is roughly a third of the critical value for which the universe neither expands forever nor collapses in upon itself.

Cosmologists use the terms "open" for a universe that expands forever, "closed" for a universe that ultimately collapses, or "flat" for one that is balanced between the two. These terms are from the vocabulary that describes the mathematical curvature of the universe. The mass density of the universe, in principle, could be any value. It is interesting, then, that our measurement is so close to the critical value for which the universe would be flat. It is all the more remarkable, because the standard inflationary model for the origin of the universe, which many theorists believe must be correct, predicts that the universe is precisely flat.

The curvature of the universe is also dependent on Einstein's cosmological constant; for a flat universe, the mass density and the cosmological constant *added together* equal the critical value. Therefore, the value we derive from our Sunyaev–Zel'dovich effect data may not determine the ultimate fate of the universe. As discussed above, however, the location of the peaks and valleys in the power spectrum of the cosmic microwave background anisotropy (Figure 2) can be used to measure curvature of the universe, and therefore determine a combination of the mass density and Einstein's cosmological constant. With our Sunyaev–Zel'dovich-effect-derived mass density and the power-spectrum-derived value (which we hope to determine in the next few years), we may be on the verge of showing that the cosmological constant is far from being Einstein's greatest blunder, as he referred to it, but is instead perhaps one of his most important contributions to modern cosmology.

We plan to continue improving our Sunyaev–Zel'dovich effect system and refining our analyses. My expanded group includes researchers involved in the next generation of x-ray satellites. We have only begun to scratch the surface of the deep pool of Sunyaev–Zel'dovich effect science. We are particularly excited about the possibility of using our instrument to survey large portions of the sky. With the current sensitivity of our system, such a survey will take 30 days or more to survey a square degree with a noise level of 40 μK. To obtain our goal of a 10-μK, noise level would take 16 months of solid observing. Considering that we only have one or two months of time available per year, this goal is clearly not feasible without improving our system dramatically. Such an improvement is possible, however. If we were able to correlate the full 10-GHz bandwidth of our existing receivers, we would increase the imaging speed of our system by over a factor of 12 and would be able to obtain our goal in roughly one month! Such a survey would provide a clear view of the evolution of structure in the distant universe.

Taking a Snapshot of the Early Universe

Building on the success of our Sunyaev–Zel'dovich interferometry program, my collaborators and I at the University of Chicago are building a novel interferometer for imaging cosmic microwave background anisotropy on intermediate angular scales. The instrument, the Degree Angular Scale Interferometer (DASI), consists of 13 telescopes. The most novel aspect of the instrument are the telescopes. In fact, they are not at all like conventional telescopes, and we usually refer to them as horns. An exploded view of one of the DASI "telescopes" is shown in Figure 11; the diameter of the aperture is only 20 cm. At the operating wavelength of 1 cm, the horn is sensitive to a circular patch of sky with a diameter of 3 degrees. The 13 horns are actually mounted on a single platform, which is only 1.6 m in diameter. The 78 telescope pairs provide sensitivity to ripples on the sky on angular scales from 0.20 to 1.3 degrees (the corresponding l range is 140 to 900). The DASI team is collaborating with a

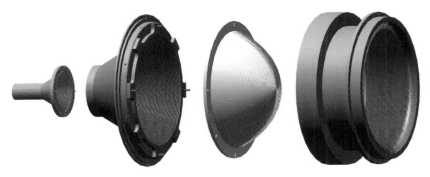

FIGURE 11   An exploded view of one "telescope" of the DASI. The telescope is unique in that there are no mirrors, and a large lens is used to improve its efficiency. The diameter of the aperture is only 20 cm, which allows simultaneous imaging of a 3-degree-wide patch of sky.

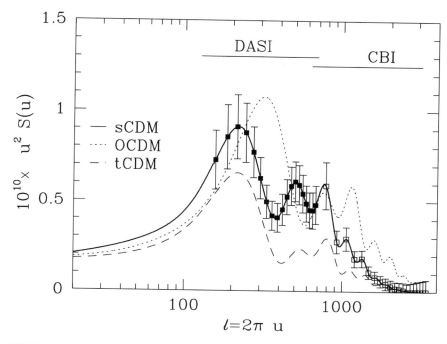

FIGURE 12   The expected sensitivity of DASI, as shown by the computed error bars, after 6 months observing from the South Pole, covering 1,000 square degrees. The solid curve is the microwave power spectrum predicted for the standard model and a flat universe, the dotted curve for an open model, and the dashed curve is the prediction for a model with a slope in the initial power spectrum of density fluctuations. DASI will be able to image one-quarter (π steradians) of the entire sky. Also shown is the expected sensitivity of the CBI for a comparable time at its planned site high in the Chilean Atacama desert.

Caltech-based team, which is building a similar interferometer, the cosmic background imager (CBI). This imager uses 1-m telescopes to cover the angular range 4–20 arcminutes (range 630–3500). We designed the instruments to provide detailed pictures of the cosmic microwave background and to characterize completely its angular power spectrum. Figure 12 shows the predicted uncertainties obtained with the instruments after 6 months of observing. Both instruments will be located at high dry sites to minimize atmospheric noise; the DASI will start observing from the South Pole in November 1999, and the CBI will begin observing from the high Chilean Atacama desert near the beginning of 1999.

    Both instruments employ the same type of low-noise receivers as used in our Sunyaev–Zel'dovich effect experiment. The large number of telescope pairs (78) and the superb atmospheric conditions allow the instruments to make high-quality images of a small portion of the sky in a single day. After several months, a detailed snapshot of the early universe and a measurement of the microwave background power spectrum will be produced (see Figure 12). This snapshot will be used to test entire classes of cosmological models. And, by simply locating the position of the first peak in the microwave background power spectrum, we will determine the sum of the mass density of the universe and Einstein's cosmological constant to better than 10 percent in only 6 months. As the instruments increase the size and quality of the snapshot, we will steadily increase the precision of all the cosmological parameters.

FUTURE PROPECTS

The promise of using the cosmic microwave background radiation to explore the early universe and to unlock the secrets of the origin of structure in the universe was immediately appreciated when Arno Penzias and Robert Wilson discovered the cosmic microwave background over 30 years ago. Since that time, theorists have shown exactly how the information can be encrypted in the background radiation as small differences in the brightness of the background from one part of the sky to another. And experiments have shown that these fluctuations in the otherwise smooth background, of an order of a few hundred-thousandths, are indeed tiny. Only now, however, are we developing sufficiently sensitive and carefully optimized instruments with which to unveil the wealth of information contained in these small fluctuations. I firmly believe that dedicated, specially designed interferometers will play a major role in this exciting field, and I plan to continue to play an active role in their development and use.

What have we learned so far? From our Sunyaev–Zel'dovich effect data we have a completely independent measurement of the expansion rate of the universe, the Hubble constant, which also gives an age of the universe of the order of 10–15 billion years. We have also used the data to measure the ratio of ordinary mass in the universe, the stuff we are made of, to the total mass of the universe. We find that of the order of 90 percent of the universe is made out of some other form of matter. Not only does the universe not revolve around us, but it is made up of completely different material.

What is the ultimate fate of the universe? Our Sunyaev–Zel'dovich effect data tell us that the mass density is about 30 percent of the critical value for which the universe balances between expanding forever or collapsing upon itself. Why is the mass density so close to the critical value, when, in principle, it could have been any value at all? Why not a billionth or 10 million? Perhaps it indicates that the inflationary model for the origin of the universe is correct, as this model predicts that the curvature of the universe is flat, which is only the case for a universe finely balanced between expanding forever or eventually collapsing.

By careful measurements of the tiny temperature fluctuations in the cosmic microwave background, we will soon know the curvature of the universe. If it is flat then we can be assured that inflation is correct. If it is flat then why do our Sunyaev–Zel'dovich effect measurements indicate that the mass density is only 30 percent of the value needed for a flat universe? A measurement of the curvature is actually a determination of the sum of the mass density and Einstein's cosmological constant. Einstein introduced a cosmological constant to his equations so they would not predict an expanding universe. After Edwin Hubble showed that the universe was indeed expanding, Einstein referred to the cosmological constant as his biggest blunder. Perhaps we will find that the universe is flat and that, far from being a blunder, the cosmological constant may be one of Einstein's many important contributions to cosmology.

In trying to answer our questions, we are led to new ones. What is the matter that makes up most of the universe? If inflation is shown to be the correct model for the early universe, then we must revise drastically our notion of the Big Bang. Inflation holds that our universe is the result of a small inflated bubble, and that countless bubbles are possible. Our understanding of our place in the universe, or even the place of the universe as we currently envision it, must once again undergo a fundamental change.

6

Looking for Dark Matter Through a Gravitational Lens: A Next-Generation Microlensing Survey

Christopher W. Stubbs

A GOLDEN AGE

It comes as something of a surprise that the luminous stars and gas that we can see and photograph are, in fact, the minority component of a typical galaxy. We now know that the beautiful textbook photographs of spiral galaxies are not accurate renditions. Most of the galaxy is made up of a substance that we cannot see and do not yet understand.

How can astrophysicists make this bold claim? Until recently, any determination of the inventory of matter in a galaxy was based on counting the objects in the galaxy that either emit or absorb light. Lately, however, techniques have been developed that allow astronomers to measure directly the mass of a galaxy. By using stars as tracers of the strength of the gravitational pull of a galaxy, we can measure the galaxy's mass.

The total mass of a typical galaxy inferred in this fashion exceeds the mass of its stars and gas by as much as a factor of 10. The rest, the galactic "dark matter," is therefore by far the gravitationally dominant constituent of galaxies, including our own Milky Way. As outlined below, understanding this dark matter is the key to resolving many of the open questions in cosmology and astrophysics.

On a much larger scale, the average density of mass in the universe determines whether the observed expansion will continue forever, or whether the Big Bang will be followed by a Big Crunch in which the uni-

Departments of Physics and Astronomy, University of Washington

verse will collapse back onto itself. There are strong theoretical reasons to believe that the universe is delicately balanced between these two possibilities, as this is the only condition that does not typically lead to either a runaway expansion or a rapid recollapse on times much shorter than the observed age of the universe. This theoretically favored condition of delicate balance between the expansion rate and the mass density is termed a "flat universe." Whether the universe has elected to conform to our preferences is, however, a question that must be settled empirically, by observation and experiment!

It is important to bear in mind that there are two distinct dark matter problems. One, the galactic dark matter problem, arises in trying to understand the observed properties of galaxies. The other, the cosmological dark matter problem, comes primarily from comparing the total observed luminous mass with the amount needed to produce a flat universe.

This is a golden age of observational cosmology. Some very basic questions have only recently passed from the realm of philosophy to scientific speculation to experimental inquiry. These questions include

- What is the eventual fate of the universe? Do we live in a flat universe?
- How did the complex structure we see today on large scales (with galaxies distributed in vast tendrils and sheets) evolve from the uniform conditions that prevailed after the Big Bang?
- What are the constituents of the universe? Is there a previously unknown family of elementary particles, generated in the cauldron of the hot Big Bang, that dominates the matter density of the cosmos, while so far eluding detection in accelerator experiments?
- What accounts for the shortfall between the total mass of a galaxy and the much smaller mass that resides in its stars and gas?

Dark matter, on both the galactic and cosmic scales, is a common thread that runs through all of the above questions. Dark matter governs the majestic rotation of the arms of spiral galaxies. It determines the interactions between neighboring groups of galaxies in the cosmos. When the universe was young, dark matter catalyzed the growth of galaxies and the evolution of large-scale structure. Finally, the overall abundance of dark matter determines the eventual fate of the universe.

Understanding the nature and distribution of dark matter are among the outstanding open questions in the physical sciences today. I am working with an international team of astronomers, physicists, and computer scientists to address these issues. We have performed an ambitious experiment using gravitational microlensing, an important new tool that is described below.

We may have detected the dark matter of the Milky Way. The present evidence is tantalizing but tenuous. Have we found the long-sought galactic dark matter? Fortunately, recent advances in technology and techniques will enable us to answer this crucial question experimentally. This will require that we mount a next-generation gravitational microlensing dark matter search. In this chapter I present the case for doing so.

This is a special time in human history. For the first time we may be able to comprehend the grand cosmological scheme that is unfolding before us. Understanding the dark matter and the role it plays is critical to our taking this intellectual step forward.

THE DARK MATTER PUZZLE

The evidence for dark matter in galaxies started to accumulate in the mid-1970s. By the following decade it became clear that essentially all galaxies, including our own Milky Way, are surrounded by extensive halos of dark matter. Just as the orbits of the planets about the solar system can be used to ascertain the mass of the Sun, astronomers can use the subtle orbital motions of stars to trace the gravitational strength of an entire galaxy. Measurements of the internal motions of many hundreds of galaxies provide incontrovertible evidence that stars, gas, and dust alone cannot account for their observed properties. Dark matter halos are thought to be roughly spherical, extending far beyond a galaxy's stellar component. Just how far these dark matter halos extend, and therefore how much total mass they contain, is a topic of considerable current debate.

Attempts to measure the distribution of mass on much larger scales, to determine the overall matter density of the universe, have proven more difficult. Recent results tend to favor the interpretation that the mass density falls short of the critical value needed for a flat universe, but uncertainties are still large. How much does the material in galaxies (including their dark matter halos) contribute to the universe's mass budget? To answer that question we will need better determinations of the amount of dark matter on both the galactic and the cosmological scales.

An important aspect of the puzzle, that branches into the realm of elementary particle physics, is the nature of the dark matter. Is the dark matter made up of some exotic elementary particle that has eluded detection in accelerator experiments, or can we construct a viable picture using the known menagerie of particles? To answer this we need to know how much ordinary matter there is.

One of the triumphs of contemporary astrophysics has been the combination of theory and observation in determining the cosmic abundance of ordinary "baryonic" matter. Baryonic matter refers to material made of neutrons and protons, the building blocks of atoms. As the universe cooled following the Big Bang, the abundances of light elements (namely, isotopes of hydrogen, helium, and lithium) were determined by the overall density of the baryons. Our theoretical understanding of nuclear physics has been brought to bear on studying the processes that governed the production of these light elements. Taking this in conjunction with recent observations of the actual cosmic abundances of these primordial elements, a consistent picture emerges in which the baryonic matter density is at most 5–10 percent of the critical value needed for a flat universe.

If observations conclusively show that the overall mass density of the universe significantly exceeds the amount thought to reside in baryons, then this would lend significant support to the idea that some new aspect of elementary particle physics must be responsible for most of the cosmological mass density. Interestingly, however, the baryonic fraction of the universe is in rough agreement with the amount of mass thought to reside in dark halos of galaxies.

A key set of measurements in the coming decade will be invaluable in sorting out this puzzle. These include (1) experiments undertaken to identify the nature and amount of dark matter in our own galaxy, (2) measurements of the evolution of the geometry of the universe, (3) studies of the large-scale structure of the distribution of galaxies, and (4) detailed maps of the apparent temperature of the cosmic microwave background, the afterglow of the Big Bang.

Each of these is an essential ingredient in understanding the amount of dark matter (on both the galactic and the cosmic scales) and how it relates to the structure of galaxies and the fate of the universe. In particular, measurements of the cosmic microwave background promise to provide information on both the overall cosmological matter density and the baryonic fraction. These will be critical pieces of information in determining how much nonbaryonic dark matter exists. Although this determination of cosmological parameters will be a watershed in observational cosmology, it will not tell us the actual composition of the dark matter.

WEAKLY INTERACTING MASSIVE PARTICLES VERSUS MASSIVE COMPACT HALO OBJECTS

There are two broad categories of dark matter candidates: astrophysical objects and elementary particles. One class of elementary particle dark matter candidates are weakly interacting massive particles (WIMPs). This play on words has its origins in the fact that these particles are thought to have interactions with matter that are governed by the "electroweak" interaction, one of the fundamental forces of particle physics. Not to be outdone, the astrophysical community has dubbed their favored class of candidates as massive compact halo objects (MACHOs).

Searching for dark matter is difficult. The only evidence we have for its existence comes from its gravitational influence on its surroundings. As far as we know it neither emits nor absorbs electromagnetic radiation, which precludes direct detection with the traditional tools of astronomy. If the dark matter is some exotic elementary particle, then in order to have eluded detection it must interact with ordinary matter very weakly, if at all.

Despite these difficulties, a number of experiments are under way to search for particular dark matter candidates. Searches for elementary particle dark matter candidates exploit some trait associated with a given hypothetical particle. One class of experiments searches for evidence of rare interactions between WIMPs and a sensitive detector. As WIMP interaction rates are expected to be in the range of a few events per kilogram of target material per day, the main experimental challenge is in understanding and overcoming naturally occurring sources of radioactive background. These can either mask or masquerade as a detection of dark matter. WIMP experiments that have the requisite sensitivity and background discrimination are now moving from the prototype stage to full-scale operation.

LOOKING THROUGH A GRAVITATIONAL LENS TO SEARCH FOR THE DARK MATTER IN THE MILKY WAY

The Principle of Gravitational Microlensing

Perhaps the most dramatic progress in dark matter searches in the past five years has been in searches for astrophysical dark matter candidates, MACHOs, that exploit the one thing we know for certain about dark matter—that it exerts a gravitational force on its surroundings.

As stressed by Paczynski (1986), if the dark matter halo of the Milky Way contains MACHOs, occasionally one will pass close to the line of sight between Earth and a distant star. The light coming from the background star will then be deflected due to the gravitational force from the intervening MACHO, as shown in Figure 1. This provides a very elegant and effective technique to search for MACHOs in the dark matter halo of the galaxy by looking for their gravitational effect on light from stars that reside beyond the halo.

The deflection of the incoming light by the gravitational field of the MACHO is just as if an optical lens of astronomical proportions had been placed between Earth and the background star, making the star appear

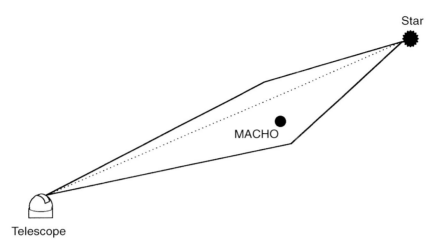

FIGURE 1   Schematic representation of gravitational microlensing. Light travel-ing from a distant star is deflected in the presence of a MACHO, which acts as a gravitational lens. The star undergoes a transient brightening as a result of the lensing. This is the basis for searches for astrophysical dark matter candidates.

brighter. As this manifestation of gravitational lensing occurs in the im-mediate vicinity of our galaxy, rather than over cosmological distances, this phenomenon is termed "gravitational microlensing."

The observable signature of gravitational microlensing relies on the fact that this precise alignment between Earth, the intervening MACHO, and a background star is fleeting. The participants in this conspiracy are in relative motion. The line of sight sweeps along due to the motion of the Earth within the galaxy, and any MACHOs in the dark matter halo would have speeds of hundreds of kilometers per hour. This means that the signature of microlensing is a transient brightening of a background star, with a very specific shape predicted by general relativity.

The duration of a microlensing event by a halo object with the mass of the planet Jupiter is expected to be about three days. The microlensing technique is therefore well suited to searching for astrophysical dark mat-ter candidates by making repeated nightly observations of stars using as-tronomical telescopes.

The effect of microlensing can be very substantial, with the apparent brightness of a star increasing by a factor of 40 or more. The duration of a microlensing event depends on a number of factors, including the mass of the intervening lens, its position, and its speed across the sky. A slow, low-mass lens can produce a signal that is indistinguishable from that of a more massive object that is moving rapidly, for example.

One great strength of this approach is that it is based on the one thing that we do know about dark matter, namely, that it exerts a gravitational influence on its surroundings. A microlensing search for dark matter is therefore sensitive to any population of astrophysical objects in the halo of the galaxy, as long as their mass lies between that of the Earth and 10 times the mass of the Sun. This broad acceptance in the experiment, over six decades in mass, is unique among dark matter searches. It nicely brackets the expected mass range of the most heavily favored astrophysical dark matter candidates.

So much for the good news. Although the signature of microlensing is unique, and the technique is sensitive to a broad range of masses, the expected rate of microlensing events is somewhat sobering. At any given time, even if the halo of our galaxy were entirely accounted for by MACHOs, only about one star in a million would be significantly brighter because of microlensing.

Implementing a Search for Gravitational Microlensing

The experimental challenge in mounting a search for dark matter using gravitational microlensing is to monitor the brightness of many tens of millions stars on a nightly basis, and then to search through the data to find the needle in the haystack: the handful of stars that brighten because of the lensing effect of an intervening MACHO. Just to make the experiment even more interesting, there are intrinsic sources of stellar variability that must be successfully discriminated against. Fortunately, there is no variable star that looks like a microlensing signal.

The ingredients for a successful search for microlensing include (1) a telescope dedicated to the endeavor to allow nightly measurements; (2) a wide-field high-sensitivity camera system for efficient measurements of many stars at a time; (3) computer resources at the scale necessary to process and store the torrential data stream; and (4) a population of millions of stars that lie beyond most of the Milky Way's dark matter halo, but that are close enough for individual stars to be resolved and measured.

A search for MACHOs using gravitational microlensing was feasible by the late 1980s because of progress in computer and detector technology. Silicon detectors, namely charge-coupled devices (CCDs), provided astronomers with a powerful alternative to film. These detectors have roughly 100 times the sensitivity of film and provide immediate digital data that are amenable to computer analysis.

At the same time, high-end computing power was becoming ever more affordable. The computing aspect of a microlensing search is daunting. There is a premium on detecting ongoing microlensing events as they are occurring. This requires first extracting the 10 million stellar images from 5 Gbytes of raw image data per night. The most recent measurement on each star must then be compared with the recorded history of prior observations. Any stars that exhibit a brightening are then subjected to further analysis and filtering.

The MACHO Project

Recognizing that technology was at hand for a microlensing search, a team of astronomers, physicists, and computer scientists from the United States and Australia banded together to undertake the MACHO Project. The objective of the endeavor was to test the hypothesis that the dark matter halo of our galaxy comprises astrophysical objects. As one of the original members of the MACHO team, I supervised the construction of what was at the time by far the largest CCD camera in the world, one that produced 77 Mbytes of data per frame. This camera is mounted on the 50-in. telescope at the Mt. Stromlo Observatory in Australia to monitor stars in the Large Magellanic Cloud (LMC), the Small Magellanic Cloud (SMC), and the center of the Milky Way. The MACHO Project started taking data in 1992 and is scheduled to cease operation at the end of 1999.

The Large and Small Magellanic Clouds are nearby galaxies, visible only from the Southern Hemisphere, which contain the millions of extragalactic stars that are needed as a backdrop for a successful microlensing search. The lines of sight toward the Magellanic Clouds form significant angles with the disk of the Milky Way, so microlensing along these lines of sight is dominated by halo objects. On the other hand, by looking toward the center of the Milky Way as well, the experiment is sensitive to microlensing by ordinary stars in the disk of the galaxy. The MACHO Project monitors stars along these three lines of sight and has now detected instances of microlensing in each of these directions.

Detection of Microlensing Events

At the time of this writing the MACHO team has amassed over 5 Tbytes of raw data and detected more than 200 candidate microlensing events, over 10 times more than any competing project's tally. The overwhelming majority of events are seen toward the galactic center, the result of microlensing by ordinary stars in the disk of the galaxy. An example of the brightening of a star in the galactic center that is due to a microlensing event is shown in Figure 2. This event was detected well before peak using the real-time data analysis capability that we developed. This allowed multiple telescopes worldwide to concentrate on the event, obtaining detailed information about the gravitational lensing system.

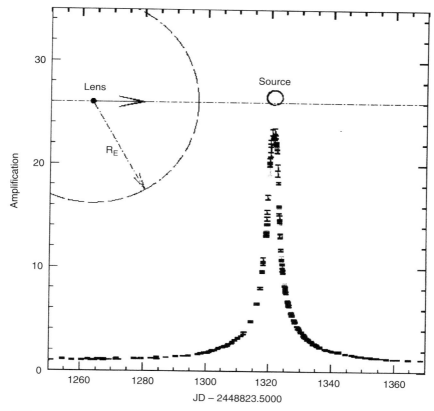

FIGURE 2   Example of a microlensing event showing the light curve of an actual microlensing event, detected in a star in the galactic center. Shown is the relative observed brightening of the star over many days. The data points are drawn from a variety of telescopes around the world. The high density of sampling allows for a detailed comparison with the predicted shape of the brightening from the theory of general relatively. SOURCE: MACHO and Gman Collaborations (1997).

This microlensing event, and hundreds like it, have shown that the experiment works. We know how to extract microlensing events as they are happening, and we communicate information about ongoing events to astrophysicists worldwide. The field of microlensing is growing rapidly. Even though the first candidate events were reported in only 1993, we have already progressed to the stage where certain exotic microlensing events lie at the intellectual frontier. For example, during a microlensing event that lasts many tens to hundreds of days, the Earth's orbit around the Sun produces a slight skew in the observed light curve. This subtle perturbation can be used to determine whether the lens is nearby or far away for a given mass of the lensing object. As described below, this effect will be critical to understanding the puzzle that is presented by the current microlensing results.

In addition to the main objective of the survey (i.e., searching for gravitational microlensing), the project has also produced an unprecedented catalog of well-sampled variable stars. This by-product dwarfs all previous surveys of stellar variability. In addition, microlensing may well provide a new window into the detection of an extrasolar planetary system. These topics are discussed briefly below before turning to the dark matter results from the MACHO Project.

Variable Stars from Microlensing Survey Data

The MACHO Project has made more sequential measurements of stellar brightness than the entire previous history of astronomy. It is therefore not surprising that the data set constitutes an unprecedented resource for the study of stellar variability. Although this was not the primary goal of the project, from the outset we knew that this would be an important contribution to astronomy. Examples of important variable star studies that have been made possible by the quality and sampling of the MACHO data include work on periodic variable stars, eclipsing binary stars, x-ray sources, and the detection of multiple novae and supernovae. We have detected many tens of thousands of variable stars, nearly all of them previously unknown.

A variety of research programs have been carried out using the MACHO Project's variable star data set. One example is a study of the detailed behavior of a class of periodic variable stars called Cepheids. These stars exhibit a tight correlation between their pulsational period and their intrinsic brightness. Measurements of the period and the apparent brightness of Cepheids can be used to extract a distance to the galaxy containing the periodic variable stars. The unprecedented temporal sampling and completeness of the MACHO data set have allowed new insight into the fundamental mechanism of pulsation in these stars. This has a direct bearing on our understanding of the cosmic distance scale.

Using Microlensing to Search for Planets around other Stars

Among the significant recent observational developments in astronomy was the announcement of the likely detection of planetary companions to a number of nearby stars. The existence of other planetary systems has long been hypothesized, of course, but only recently have we had experimental confirmation of them. The main technique that is currently used to search for extrasolar planetary systems (measuring radial velocities) is only effective for bright stars and is limited to giant Jupiter-sized planets. An Earth-like system would be undetectable using this approach.

Imagine that one of the many cases of gravitational microlensing by ordinary stars, detected toward the galactic center, was due to a star that had a planetary companion. This could perturb the shape of the detected microlensing light curve, as the light would he deflected by both the main star and its companion planet. Detailed calculations show that a significant fraction of such cases would be detectable, but that the perturbation of the light curve would last only a few hours. Furthermore, the technique is sensitive to Earth-mass planets and is unique in this regard.

Using microlensing to search for extrasolar planetary systems is in its infancy, but is an exciting and promising technique. We have identified instances in light curves of microlensing toward the galactic center that might be due to lensing by planets around distant stars, but the analysis of these events is still in progress.

Microlensing Toward the Galactic Center—A New Probe of Galactic Structure

By mapping out the event rates toward the galactic center, along different lines of sight through the disk of the Milky Way, microlensing can be used as a powerful new probe of galactic structure. Initial microlensing rates reported by both the MACHO collaboration and a competing team (the Optical Gravitational Lensing Experiment survey) both exceeded earlier predictions. These early theoretical predictions made the unwarranted assumption that the Milky Way is symmetrical, shaped like a dish. Once the microlensing results were announced, astronomers soon realized that a more realistic model was needed, specifically one that includes an elliptical body of stars in the center of the galaxy. After including the contribution of this "galactic bar," the models were brought into agreement with the microlensing data.

The constraints on galactic structure that are provided by microlensing have an indirect impact on our understanding of the Milky Way's dark matter. By probing directly the mass distribution in the disk of the galaxy, we can better understand the interplay between dark matter and stars in supporting the observed overall rotation of the galaxy. This in turn can be used to constrain models of the dark matter halo.

Microlensing Toward the Magellanic Clouds—Testing the MACHO Hypothesis

Although the lines of sight through the disk toward the galactic center have produced the overwhelming majority of the microlensing events observed to date, a handful of events seen toward the Magellanic Clouds may be the key to understanding the nature of the Milky Way's dark matter halo. Because microlensing toward the Magellanic Clouds would be dominated by MACHOs in the halo, this is the most sensitive way to detect and constrain their existence.

Over the course of two years of observations of stars in the LMC, our MACHO team produced four of the most stringent results to date on the possibility that astrophysical objects, rather than elementary particles, might make up the dark matter halo of our galaxy. We have carried out over 1,000 observations of more than 9 million stars in the LMC. This data set has led to two major dark matter results, described below.

No Low-Mass MACHOs. It is often the case in science that tremendous progress is made by eliminating possibilities. The search for dark matter is no exception. In our LMC data set we have searched for microlensing events with a duration ranging from hours to 200 days. We see no LMC events that last less than 20 days.

Using the connection between lens mass and event duration we can exclude astrophysical objects with masses between one-millionth to one-tenth of the mass of the Sun as making up the dark halo of the galaxy. The most favored astrophysical dark matter candidates, brown dwarfs and Jupiter-sized objects, fall squarely within this excluded range. These were thought to be prime dark matter candidates as they are not massive enough for their internal (gravitationally produced) pressure to ignite the nuclear burning that makes stars shine. This was thought to be a natural explanation for their having evaded direct detection. Evidently, based on the microlensing results, this speculation was not correct. Excluding such a broad range in dark matter candidates, with high statistical confidence, is a major step forward in dark matter science. If this were the only scientific result from the MACHO Project it would be rightfully regarded as a very successful endeavor.

An Excess of Long-Duration Events—The Detection of the Galactic Dark Matter? We did, however, detect eight microlensing events toward the LMC, when only about one event was expected from lensing by known stellar populations. These events typically last 80 days, corresponding to MACHOs with a few tenths of a solar mass, although the uncertainty in the mass is quite large.

Is this the long-sought galactic dark matter? Based on models of the structure of the galaxy and its dark halo that were popular before the microlensing results were announced, the answer would seem to be yes. Taken at face value, the event rate corresponds to our having detected at least half of the galaxy's dark matter halo! Should this prove to be the case, one of the major contemporary astrophysical and cosmological puzzles will have been solved.

If we have seen MACHOs in the halo of the galaxy, what could they be? Much of the power of the microlensing technique is its insensitivity to the detailed nature and structure of the lensing object; however, at this stage that becomes a disadvantage. Speculation about the composition of the MACHOs is therefore based on other knowledge of the different species in the astrophysical zoo, but it is entertaining and even instructive to contemplate the various possibilities.

One alternative is that the MACHOs are the endpoint of stellar evolution from some very early population of stars, perhaps white dwarf stars that have exhausted their nuclear fuel, or even neutron stars. A problem with this scenario is that during their evolution such stars would have suffered significant mass loss, blowing off major amounts of material that contain heavy elements. There are stringent limits on the amount of interstellar matter, and scenarios that invoke an early population of stars have difficulty in confronting these observational constraints.

Another possibility is that MACHOs are some gravitationally bound state of matter that never passed through a stellar phase. This somewhat exotic class of objects includes "primordial black holes" that were formed early in the history of the universe and have been darkly coasting along ever since. Such objects would have been virtually impossible to detect by looking for any emitted light, but they would act as perfectly decent gravitational lenses. In this case, the MACHOs might be exempt from constraints imposed on the baryonic content of the universe, as they could in principle have been formed in an era where those bounds simply would not apply.

There are, however, other possible interpretations of the observed microlensing event rate that do not invoke the dark matter halo. Particularly when we are dealing with only a few detected events, the statistical uncertainties are still fairly large. Possibilities for accounting for the excess lensing events with ordinary stars rather than halo MACHOs include

- lensing by foreground LMC stars,
- lensing by some foreground dwarf galaxy or stellar debris from the LMC, or
- lensing by some previously unappreciated extended stellar population of our own galaxy.

These are testable hypotheses, and it is imperative that we distinguish between these alternatives and the dark matter interpretation of the observations.

THE PATH FORWARD:
A NEXT-GENERATION MICROLENSING SURVEY
Why Do Another Experiment?

The compelling motivation for a next-generation microlensing survey is to determine whether the excess of events seen toward the Magellanic Clouds is due to the dark matter halo of the Milky Way. Why do we need a new experiment to answer this question?

Based on our current event rate, by the end of the 1999 the MACHO Project will likely have detected about two dozen microlensing events toward the LMC. Perhaps three or four will be seen toward the SMC. Although this significantly exceeds the rate expected from known stellar populations, it is simply too few to definitively test the idea that the lensing is due to MACHOs in the halo of our galaxy.

There are three reasons to increase the number of events toward the LMC and SMC:

• We can address the hypothesis of lensing by foreground LMC stars by mapping out the event rate across the face of the LMC. A detailed study of how the event rate depends on the density of LMC stars will allow a definitive test for this possibility. This will require many tens of LMC events to achieve the requisite statistical significance.

• A comparison of the event rates between the LMC and the SMC lines of sight probes the flattening of the lensing population. A disklike distribution of MACHOs would have a very different ratio of rates than would a halo population. To make a definitive comparison will require hundreds of events.

• Finally, about 10 percent of the events that we have detected to date show structure that cannot be accounted for by the simple point-mass, point-source picture given in Figure 1. These exotic lensing events are now well understood and are very useful. In these cases the degeneracy between lens position, mass, and velocity breaks down, and we can learn a great deal about the lensing system. In particular, for long events the orbital motion of the Earth introduces a measurable perturbation that is very different for nearby lenses versus distant lenses. This is a very powerful way to establish whether the lensing population is in the immediate foreground (perhaps a wisp of stars from the galaxy) or is far away (as expected for halo MACHOs). Taking full advantage of this phenomenon will also require the detection of roughly 100 events.

We need between 100 and 200 Magellanic cloud events to achieve these important goals. The present experiment would need to run for decades to detect the requisite number. Other ongoing microlensing searches have an apparatus that is essentially similar to that of the present MACHO Project. Once they come into full operation they will have roughly comparable event rates. A new experiment is necessary, building on the successes of the existing program, to achieve these objectives.

There are two possible answers to the experimental question of whether the excess event rate we have seen is due to dark matter MACHOs. It is worth stressing that with either outcome this will be a crucial and successful experiment. If the results support the halo MACHO hypothesis, then one piece of the dark matter puzzle will be firmly in place. On the other hand, if further intense scrutiny shows that the detected signal is due to some previously unappreciated ordinary stellar population, then microlensing searches will have essentially eliminated astrophysical objects as viable dark matter candidates. This would also be a great step forward in unraveling the mystery of dark matter. We will win in either case.

The Plan

Fortunately a revolution in technology is not necessary to mount a search that would detect events at well over 10 times the current rate. We will need to monitor over 10 times as many stars per night than are scanned by the existing MACHO program. A number of performance enhancements relative to the MACHO Project will make this possible, including

- doubling the camera's sensitivity by using cutting-edge CCD technology;
- doubling the camera's field of view to one full square degree;
- halving the time spent reading out the CCD camera, thereby increasing efficiency;
- more than tripling the experiment's light-gathering power, with a 2.4-m telescope;
- more than tripling the number of distinct stars per frame by operating at a site with minimal atmospheric degradation of image quality;
- reducing the background in the images by operating at a darker site; and
- establishing very close coordination between the survey and the network of follow-up telescopes.

The next-generation microlensing survey should easily surpass the total number of Magellanic cloud events detected in the past five years during its first year of full operation.

The challenges in carrying out this program include the fabrication of the new instrument and obtaining access to the requisite (large!) amount of telescope time at an outstanding astronomical site. In addition, the raw data rate and database management problems will be an order of magnitude more difficult to handle than in the current microlensing survey. Although demanding, these are all manageable problems given the right combination of personnel, resources, and contemporary technology. This convergence is now taking place, and a capable team of scientists is laying the necessary groundwork for the next-generation experiment.

The Site and the Telescope

Turbulence in the atmosphere degrades astronomical images. The superb resolution obtained by the Hubble Space Telescope is possible only because it orbits above the swirling layers of air that plague ground-based observatories. Unfortunately the field of view of the Hubble is far too small to carry out a search for microlensing. The penalty extracted by the atmosphere depends very much on the location of the observing site. Ideal sites for optical observatories are in high, dry places where the topography rises rapidly above the ocean or a plain.

A longstanding difficulty in designing, building, and operating telescopes for optical astronomy has been to actually realize the image quality potential of a site. There has been much recent progress in understanding the aspects of telescope design that influence image quality. A new generation of telescopes are being engineered with the express goal of delivering images that are limited only by the characteristics of the site, not the telescope.

The performance of the next-generation microlensing survey will depend critically on the image quality delivered to the camera system. The best astronomical sites in the Southern Hemisphere are in Chile. The Cerro Tololo Interamerican Observatory (CTIO) is a U.S. federally funded facility in Chile that supports astronomy in the southern sky. Tests of the stability of the upper atmosphere in Chile indicate that exquisite image quality should be achievable there.

The next-generation microlensing team is working in tandem with the National Optical Astronomical Observatories (of which CTIO is a part) to construct a new technology telescope in Chile with an aperture of 2.5-m. During the times when the Magellanic Clouds are visible, the telescope will be devoted to the microlensing survey until the target number of events are detected. Present plans place this telescope on the same ridge that was selected for the southern Gemini telescope (the flagship 8-m-diameter U.S. telescope that is now under construction).

The Camera System

The construction of the instrument described above is a serious technical challenge. The CCD detectors used in contemporary astronomical instruments are the largest integrated circuits made, and they are remarkable devices. They convert incident light into an electrical charge, in proportion to the light intensity. They perform this conversion of light into an electrical signal with an efficiency of over 90 percent. The resulting charge can be determined at the level of a few electrons. Individual detectors with 2048 × 4096 pixel formats are becoming common.

In the camera being proposed for the next-generation microlensing survey, 18 such detectors must be carefully aligned within a vacuum system. They are then cooled to about −100 degrees C in order to suppress background signals that are thermal in origin. This focal plane mosaic array of detectors must be interrogated by a readout system that converts the light-generated electrical charge into digital data that are suitable for computerized data analysis. These data will typically flow to the analysis computers at the rate of 20 Gbytes per night.

A number of innovative features will be incorporated into the instrument. For example, any image degradation from slight vibrations and motions of the telescope will be suppressed by making compensating shifts in the charge on the CCD detectors while the image is being exposed. Also, a different subsystem will continually ensure that the image is kept in focus. This is consistent with the survey-wide emphasis on image quality.

Although this camera will be larger, more sensitive, and more efficient than any existing astronomical instrument, it is a natural next step in the evolution of astronomical instrumentation for wide-field imaging. The camera system is one of the critical technological advances that will enable the next-generation microlensing survey to go forward.

Computing

There are two main challenges for data analysis and data management. The first is installing sufficient brute-force computing power to carry out real-time analysis of the raw image data. The second is devising a data storage system that efficiently allows scientists access to the accumulated sequence of stellar observations.

The online data storage task will likely require over a terabyte of magnetic disk. This is over 1,000 times more than the 1-Gbyte drives that are shipped with typical personal computers. Interestingly, making backup copies of data sets of this size is one of the major challenges in carrying out the project.

A networked array of 20 high-end PC-type computers is adequate for the experiment. By allocating one CPU to each of the 18 detectors in the camera, the system is well suited to taking a parallel approach to the computational task.

Timetable

Current plans call for the installation of the telescope at CTIO in 2001. Camera fabrication will occur during calendar years 2000 and 2001. After six months of shakedown and integration, the next-generation microlensing survey will commence in earnest in late 2001 or early 2002.

SUMMARY AND CONCLUSIONS

The dark matter problem is one of the pivotal open questions in the physical sciences today. We know that most of our galaxy is made of dark matter. We know that the average cosmic density of dark matter determines the eventual fate of the universe. We do not know what dark matter is or how it is distributed. This has spurred a number of efforts to try to detect either elementary particle or astrophysical dark matter candidates.

A new tool in this quest, gravitational microlensing, has produced two important results to date that bear on astrophysical dark matter candidates. First, the lack of any short-duration microlensing events has eliminated brown dwarfs and similar objects from contributing in any significant way to the dark halo of the Milky Way.

We do, however, see more long-duration events than one would expect from the conventional picture of how stars are distributed in the galaxy. This could be our first hint of an actual detection of the dark halo of the galaxy. Microlensing is the only dark matter search technique that has produced a robust and persistent signal. With only a handful of events, however, it is difficult to distinguish between this interpretation and more conventional explanations. Doing so will require a tenfold increase in the detection rate.

A next-generation microlensing survey, conducted with state-of-the-art instrumentation at an outstanding astronomical site, would provide the number of events needed to test whether we have detected the dark matter. The case for performing this experiment is compelling, as it could provide us with the key to understanding some of the most profound questions mankind has considered.

ACKNOWLEDGMENTS

I am fortunate to have had the opportunity to work on the MACHO Project, and I thank my colleagues for a stimulating and rewarding experience. In particular I am indebted to them for their friendship over the course of the project. Bernard Sadoulet, the director of the Center for Particle Astrophysics, has provided me with tremendous opportunities for which I am most grateful. Thanks also to Craig Hogan and Carrington Gregory for their comments on the manuscript. This chapter is dedicated to the memory of Alex Rodgers, past director of the Mt. Stromlo Observatory, without whom the MACHO Project simply would not have been possible. We miss you, Alex.

REFERENCES

MACHO and Gman Collaborations. 1997. MACHO Alert 95-30: First real-time observation of extended source effects on gravitational microlensing. *Astrophysical Journal* 491:436–451.

Paczynski, B. 1986. Gravitational microlensing by the galactic halo. *Astrophysical Journal* 304:1–5.

7

Advancing Our Knowledge of the Universe

David Wilkinson

HURRAY FOR PLODDERS

James S. McDonnell, who graduated from Princeton in 1921, majored in physics and is certainly one of the great Americans of this century. John McDonnell talked about his father's ability to plod. That is, he did not avoid the hard work of taking care of details. I believe that John Carlstrom and Christopher Stubbs resonate with that characterization, because experimental physicists are plodders. Most of the time, physicists do fairly mundane things. The big ideas come quickly, often when least expected, but then begins the real work—mechanical and electrical design, overseeing fabrication (often doing much of it themselves), troubleshooting, long hours of data taking and analysis—things that they enjoy doing, but which do not attract a lot of attention. John Carlstrom and Christopher Stubbs are plodders. Like Mr. Mac, their work style is *hands on*, taking care of the details.

Coming from Princeton, which is a center for astrophysics and cosmology research, I often hear talks by leading people in the field as they visit the Institute for Advanced Study or the Astrophysical Sciences or Physics departments at Princeton. The chapters by Carlstrom and Stubbs are two of the most exciting essays that I have read in the past year. Important discoveries in astrophysics and cosmology almost always accompany large advances in sensitivity, resolution, sky coverage, and observing techniques, the things that Carlstrom and Stubbs are planning to

Physics Department, Princeton University

improve. They are very likely to substantially advance our understanding of the universe we live in. As Mr. Mac demonstrated, plodders can accomplish great things.

OUR STRANGE UNIVERSE

Carlstrom and Stubbs described a lot of what we already know about our universe, but here I will add a few things that may help to put their work into context. Figure 1 shows a slice through the universe as seen by observers at the vertex. In viewing this figure we need to remember that the speed of light is not infinite. There is a delay between when light is emitted and when we see it. Therefore, when looking out into deep space, we are also looking far back in time. In a sense, telescopes are time machines, allowing us to look into the past. The other remarkable fact about

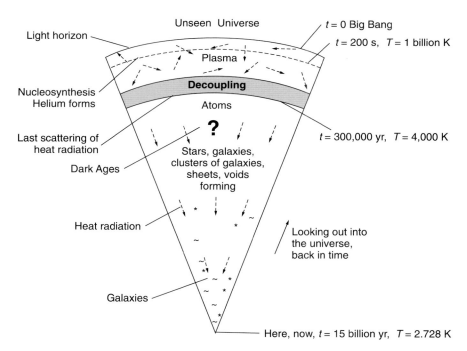

FIGURE 1 A slice of our universe showing some significant epochs in its evolution. The story is based on the big bang cosmological model, now well established by observations.

our universe is its general expansion. To us (and to everyone else in the universe) it appears that other galaxies are moving away—the greater the distance, the faster the recession speed. Curiously, because the expansion is linear (recession speed is proportional to distance), *observers everywhere* in this universe see other galaxies receding from themselves in all directions, giving the impression that they are *at the center*. Only two universes have this property; the other one is static and unstable. We do not understand why our universe has these peculiar properties; they seem to have been built in from the beginning.

Imagine that we are at the vertex of the slice looking out into the universe and back in time. On the right-hand side of Figure 1 the age of the universe is shown at the time that the light we see was emitted, the age of the galaxy as it appears to us. The nearby universe is about 15 billion years old and contains thousands of majestic galaxies with beautiful shapes and impressive complexity. Our most powerful telescopes can detect very distant galaxies as they were only a few billion years after the origin of the universe, the "big bang." Thus, the universal time delay lets us study the evolution of the universe since very early times. The scale and complexity of our universe is mind-boggling. There are roughly 100 billion galaxies in our visible universe, and each galaxy is made up of about 100 billion stars. It is an astonishing universe that we live in, and we have only just begun to understand it.

When thinking about our place in this vast, inhospitable universe, I resonate with Mr. Mac's use of the term "spaceship Earth." We live in a universe of enormous scale and complexity, and as far as we know, Earth is the only place hospitable to life as we know it. Our Earth *is* a spaceship, and there is no place else to go at a cost that we can afford. We had better take very good care of this place!

In Figure 1, the large question mark signifies what cosmologists call the "dark ages." Somehow the beautiful structure in our universe—galaxies, clusters of galaxies, huge voids, and sheets of galaxies—started to form in this epoch. Did the stars form first and then later gather to form galaxies? Or did the matter clump on galactic scales before stars started to burn? What were the first stars like? What role was played by the massive black holes we now find at the centers of galaxies? These questions are currently at the forefront of cosmology research. This whole mysterious epoch of cosmic time is just out of the reach of our most powerful telescopes. We are now observing youthful galaxies, like the ones we see in the Hubble Space Telescope's marvelous "deep-field" picture. Perhaps the next generation of telescopes, now on the drawing boards, will penetrate the cosmic dark ages.

Ironically, we know more about our universe at epochs further away in space and time than the cosmic dark ages. In his chapter, John

Carlstrom discusses the "cosmic microwave background radiation" (heat radiation) that comes to us from an epoch when the universe was so hot that energetic particle collisions tore electrons off of atoms, keeping the material ionized. The universe was filled with uniform, hot "plasma" of negative electrons and positively charged atomic nuclei. How did we get from this relatively simple plasma universe of elementary particles and heat radiation to the very complex universe we see today? We do not know. But the work that Carlstrom discusses is very likely to unveil a good deal of what went on in these cosmic dark ages. Let me explain how we can learn about the formation of the cosmic structure billions of years ago by measuring fluctuations in the cosmic microwave radiation temperature now. In Carlstrom's chapter, he mentions that the cosmic microwave radiation is coming straight to us from a time just before the dark ages, when the radiation and the matter interact for the last time. In Figure 1 it is called the "decoupling" epoch, about 300,000 years after the big bang. Before decoupling, the heat radiation was scattered around strongly by the plasma. (Note the random directions of the radiation prior to decoupling.) However, at decoupling, something very special happened. The universe had cooled (due to the universal expansion) to a temperature of about 4,000 degrees, cool enough that the electrons could now stick onto the protons and form neutral hydrogen for the first time. Once the matter in the universe was electrically neutral, gravitational forces (always attractive) began to pull the matter into clumps wherever the density happened to be a bit higher than average. Inexorable gravity continued to tighten the clumps, forming the wonderful panorama of structure that we see in the universe today. By mapping the tiny temperature fluctuations in the cosmic microwave radiation at the time of decoupling, we are taking pictures of the density fluctuations that are the seeds of cosmic structure formation. It has taken 35 years to develop the technology and the know-how to make these observations possible, but most cosmologists believe that measurements like those described in Carlstrom's chapter will make a huge advance in our understanding of the universe.

There are some other interesting features in the cosmic diagram in Figure 1. The epoch in the diagram called "nucleosynthesis" happened only a few minutes after the big bang. The temperature of the universe (a few billion degrees) and the density were just right for nuclear reactions to form helium nuclei from the protons and neutrons in this very hot plasma. Calculations show that about 24 percent of the matter should form into helium nuclei, the rest being hydrogen nuclei—protons. Remarkably, the Sun is composed of about 24 percent helium and so is every other sample of cosmic matter that we have been able to measure. It looks like the whole universe is composed of 24 percent helium in accord with the big bang primordial nucleosynthesis model. This agreement of observations and theory, based on the big bang model, argues that cosmologists are on the right track.

Finally, why is the cosmic diagram terminated in a circle labeled big bang? That is as far away as we can see, about 15 billion light years in all directions. In the 15 billion years that the universe has existed, light has not had time to reach us from outside of that boundary, which is called the "light horizon." Of course, as time goes on we see farther and farther into the "unseen universe" as our light horizon expands. What is beyond our light horizon? Probably the same stuff that is inside, but we are not sure of that. We cannot see out there directly. However, our imaginations can take us there via what Einstein called a "gedanken" experiment. Imagine an observer on another galaxy far away from us. When that observer draws her light horizon circle, it goes outside of ours in the direction opposite from us. In that direction, light from outside our light horizon has already reached her position. If the stuff between our horizon and her horizon were different, she would see a lopsided universe, different in the directions toward and away from us. If we believe that the universe is homogeneous and isotropic (the same in all directions) for observers anywhere in the universe, the unseen (by us) universe must be similar to the part that we see. Otherwise, there is something very special about our placement in the universe, contrary to observations showing us on a very ordinary planet, orbiting a very ordinary star, far from the center of our galaxy.

HOW DO WE KNOW THAT?

So that is the big picture. Why should we believe such a fantastic story? I like to teach courses for nonscientists, and I always try to get them to ask the question: *How do we know that?* It is the most important question in science. How do we know (never for certain, but with a high confidence level) that the essential parts of the big bang story are true? Primordial nucleosynthesis is a good start. Astronomers have trouble explaining the observed abundance of low-mass elements, like helium, except by production in a very early stage of the hot big bang. I am particularly impressed that the physics that we have learned in our tiny corner of the universe can explain something as grand and remote as primordial nucleosynthesis.

Additional compelling evidence that we live in a big bang universe comes from a long series of observations of the spectrum of the cosmic background radiation. A spectrum measures radiation intensity at different wavelengths, as illustrated in Figure 2. George Gamow predicted the spectrum of the cosmic microwave background radiation in the late 1940s,

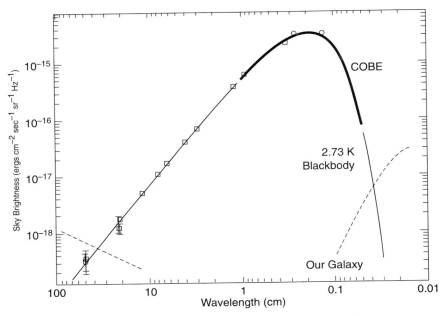

FIGURE 2 Cosmic Blackbody Radiation Spectrum. The results of many measurements of the spectrum of cosmic microwave background radiation. The thin solid curve is the theoretical prediction, a blackbody curve. The thick curve near the peak is the result from the Cosmic Background Explorer (COBE) satellite. The experimental errors on the COBE measurement are smaller than the width of the thick curve. The squares and circles are results from ground- and balloon-based measurements. Only the big bang model accounts for such accurate agreement between theory and measurement.

15 years before the radiation was discovered. Using the very hypothetical (at that time) big bang model, Gamow and his collaborators, Ralph Alpher and Robert Herman, found that the heat radiation remnant of a hot early universe should have an almost perfect "blackbody" spectrum. This is a spectrum that every undergraduate physics student studies because it describes a very common and simple phenomenon. All hot bodies that are good absorbers of radiation emit heatlike radiation with a blackbody spectrum, which is completely characterized by the body's temperature. (Sunlight is part of the Sun's blackbody spectrum emitted by its 6,000-degree surface.) Interestingly, the blackbody spectrum played an important role in the discovery of quantum mechanics early in the twentieth century. Max Planck had to hypothesize quanta of radiation to concoct a theory explaining the peaked shape of the blackbody spectrum emitted by hot bodies. A critical test of the big bang cosmological model was to measure the spectrum of cosmic microwave background radiation to see

if it has the predicted blackbody shape. The measurements spanned 25 years, culminating in a very accurate measurement by NASA's Cosmic Background Explorer (COBE) satellite in 1990. The data are shown in Figure 2 along with the theoretical blackbody curve. As can be seen, the data fit the theory extremely well, especially in the critical region of the peak of the curve. Note that the curve peaks at about a 2-mm wavelength, in the microwave band. This means that the current temperature of the radiation is a few degrees above absolute zero. The enormous expansion of the universe has cooled the radiation from billions of degrees at the time of nucleosynthesis to a few degrees at the current epoch. (The temperatures at several cosmic epochs are shown on the right-hand side of Figure 1. The unit of temperature is the Kelvin, or K, in degrees above absolute zero.) The COBE measurement gives a cosmic radiation temperature now of 2.728 ± 0.004 K. It is important to remember that the spectrum had been predicted *before the radiation was discovered*, using the big bang model. Furthermore, no other cosmological model comes close to explaining the measured blackbody spectrum. The agreement between theory and accurate measurements of the spectrum makes a very strong case for the big bang cosmological model.

There are two other interesting subplots of this story. I mentioned above that the blackbody spectrum was a cornerstone of the discovery of quantum mechanics. Now we find that blackbody radiation fills our universe and played a central role in the discovery that we live in a big bang universe. I doubt if any other physical phenomenon will be as important to our understanding of physics and cosmology over such an enormous range of scales. Second, in Figure 2, note that there are some dotted curves showing emission from our own Milky Way galaxy, radio waves on the left and emission from galactic dust on the right. Further to the left and right, these emissions from our galaxy rise to completely overwhelm the cosmic microwave background radiation. How remarkable that the heat radiation from the big bang happens to be centered in an excellent wavelength window for seeing out of our galaxy. There is no evolutionary reason for this; it is just *pure luck*. To observe the seeds of structure embedded in the cosmic microwave background, we need to measure intensity fluctuations with an accuracy of 1 part per million. I cannot believe how lucky we are that right at the peak intensity of the cosmic blackbody radiation our galaxy's emission is about a million times weaker!

FINAL REMARKS

Christopher Stubbs and John Carlstrom are working on two of the most important problems in cosmology today. We live in a universe apparently dominated by dark matter about whose composition and form we know very little. There are so many possibilities—unknown elementary particles, black holes, dim or dark stars, neutrinos with mass—that we do not know where or how to look for the dark matter. One needs to try all reasonable and testable ideas. Christopher Stubbs is looking in the halos of galaxies, where many people think the dark matter resides. This is a very difficult problem.

Questions about the past and future evolution of our universe are better defined; we have a better knowledge base to build on. Studying the seeds of structure formation by measuring tiny fluctuations in the cosmic microwave background radiation is a very difficult technical problem, but at least one knows that this is the right place to look. John Carlstrom may not find what he expects, but he is bound to find something very interesting about the universe. Mapping the cosmic microwave fluctuations will give a big boost to our ongoing efforts to measure important cosmological parameters—the numbers that define the kind of universe that we live in. For example, the parameters will tell us whether we live in a universe that will expand forever, becoming more rarified and cold, or eventually stop and recollapse to a hot, highly compressed state. The cosmological parameters are being measured by combining the results of many different kinds of astronomical measurements. John Carlstrom's work on the microwave background radiation fluctuations and on the Sunyaev–Zel'dovich effect in clusters of galaxies are two important pieces to the cosmic puzzle. I look forward eagerly to following the progress of these two very talented experimental physicists in the years ahead.

8

Putting the Genome on the Map

Wendy Bickmore

Imagine an essay 5,000 times longer than this one. You probably would not be able to pick it up, let alone want to read it, because you would have trouble finding the 80,000 words embedded within its seemingly unintelligible list of 3,000 million characters. This is the scale of the human genome "map," an international project to determine the sequence of bases (characters)—A, C, G, or T—of DNA that are strung out along the 23 pairs of our chromosomes and to find the position of the genes (words) within this list. The genes are the parts of the sequence that encode the instructions for making all of the proteins necessary for life, growth, and development. Storing, accessing, and making any sense out of a map of these proportions will require considerable ingenuity, but the completed human genome map will be a great achievement and will make an unparalleled contribution to our understanding of normal biology as well as to our understanding of disease.

Most of us are familiar with the concept of maps in our everyday lives and the diversity of information that they can, and should, communicate. They enable us to navigate our way around new and often complex landscapes. Hence, a good map is more than just a list of place names (or bases). A good map conveys information about different types of environment, about topological features of the landscape, and about its important structural and functional features. Such maps, full of colors and shapes, are a pleasure to behold and can conjure up an immediate mental

Medical Research Council, Human Genetics Unit, Edinburgh, Scotland picture of what the landscape really looks like. If the map of the human genome were to include this richness of information, in addition to the raw DNA sequence (Figure 1), how much more would we know about our genomes and the way that they work?

Our genetic inheritance is more than just the DNA that we get from our parents. Genetic information is encoded as DNA intimately wrapped up with many layers of protein in the form of microscopic objects: chromosomes. Gene sequence spells out the raw ingredients of the recipe to make protein, but the chromosome itself dictates whether the page of the recipe book is opened to be read in the first place.

The average human chromosome contains 100 million base pairs of DNA that would be 5 cm long if stretched out. To fit the 2-m worth of DNA from all our chromosomes into the nucleus of the cell, which typically measures only one-thousandth of a centimeter across, the DNA must be folded up tightly and precisely. This is brought about by the DNA interacting with specific proteins to form chromatin—the stuff that chromosomes are made of. Two-thirds of the mass of the chromosome is made up from proteins—the DNA that is usually the focus of so much of our attention in current maps is a minority component. Information is embedded within the details of this packaging and can modify how genes are expressed, how the genome is copied, and how it is segregated to new cells. A more complete map of the genome needs to be able to describe these facets of the chromosome.

GENES IN AN OCEAN OF DNA

Developing a more functional map of the genome has been the objective of my research for the past six years. Ironically, interest in this goal was sparked by my own efforts to develop a conventional one-dimensional map of a small part of the human genome. I wanted to use this to identify the position of genes involved in the development of a particular genetic disorder.

Genes account for only a small proportion of our DNA sequence, and there can be extremely long stretches of DNA between genes. Locating the position of genes within this sea of DNA can feel like trying to find the proverbial needle in a haystack. Fortunately, in the genomes of mammals, half of the needles have beacons on them—termed CpG islands—which allow us to home in on the genes. CpG islands are found at the beginning of 60 percent of our genes, and they have sequence characteristics that make them stand out from bulk DNA. DNA-cutting enzymes that recognize the types of sequence found in CpG islands can therefore be used to pinpoint the position of some genes. What we noticed about the distribution of these cutting sites in our simple map was that they

```
TAAAACAAGTTTCTGTTTAAAGGGAGGAGGAAGCTGGAGCAGGAGGAAGAGAGCAAGAGCCAC
ACGTACTGCCCAGTAAATTAGATTCCTTAGATGATAATATGATTACTGGCTTGTTGTACAGCT
ATTTGAAATGACACTCAAATAATACACAGTATATTTTAAATCAATTAATTATTTAGGGCTTTTC
CCTCCTATTAATTTGTATATGGAAAAGCTTTATTAAACCATTTTATAAGTTAGCCCTCA
GAGTAAAATAAATTTTTAGAAATATTGTAGAAATTAACGATGCATTATCTTCCAGATT
GTGCCTATTCTCAATAAAAAAGCAAATACTAAATCAATTCCACCCACCTGGTAGCACAG
ATCCAGCGGTAGGGCTCCTCCCAGTCTGGAGCGGCTTTATTTGCATAAAGTAGTTAAGG
GGAGAGAGGAAAAGCGACAGCAGGCCGGAGGGGAAAACATACTTTGCATAAAGTAGTTAAGG
TTCAAGCCAAGAGCCGGATTTCTCATCTTTCTCTCTCCTACATACTTCCCCTACTTCAAG
GCCTCTACGCCCATCCAGCGTCTCCACCCAGCCTCTCGGCTGATGGGGAAGCTCTGG
GGTGGAGAATACACTCAAATTATAATGGTCCTAGATCGCAAACAATTCCGCCTCACTGTGC
CGGGCTGAAGGAGGCCTGAAAGAAAAGGTTTCTTCCCGGCGTCTCCAAGAATCAGTTGCAG
CCAGCTACAGCCTGCCTCAGTAGCACTCGGCATTTCGGGCTCTACGAGTGTGCCTAGAGAG
GCCCCGGCCGCCCGGGTACCCTTCTTCCCGATGTCTCCGGCCTCCAGAATCCAGTGAGCAG
CCTTGCCCGGACAAATGGAGAGGCGTCGTTGGCGGCCTCGGGCGGTCGGAGGGCCTAG
TTCCGGGCTCGGCTCGACCACCACCGGCGTCCGGCCAGCCCGAACCGGCCCTTTATTGACA
GCGACGATAAGCAGGCGGGGCCGTGGCGTTTCCAGAGTCGCTCTGACCACCCTGCTCGG
TTGCAGGGATTTGCGCACACTTAATGGCCTGGACTCCAAACAAGCCCATAACAGCCCTCT
GAAGGCCTACACGGTTCGGAGGCTATCCGCAGCCCACACGCCCAAGAAACTGGCCAGCTAG
CTGGGAGACCTCAGGACATTCGAGGCTGGAGCTGGGCCAGGTCGTGAGGGGGAGCTTAGG
GGCTTACCCAGAAACTAGGTCACTAGTGACATGGCGACGGGCAGCAGGGGAAACCAGAACAAA
TGCCTCTTAGCATCCCTTCCCAGCCGACGAGGGCCACAGACCATGGTAGGTCTAGG
CCGAGCTAGAGTGGCCAGTGGAGGTGGGCGCTCCTAGGCCTTAACAGGATGCCCAG
GGAAGCTTGTAGGAGAAGAAGAAGTTTGGAGATAGTTTGTAAATAAGTAGGAACATCCTGC
GGAACTTCTAGCCAAAGCTCCCATTCTCTCCCCACCTTCACCCCTCACCGCACCCCCACCTCT
AAAAAATCAAGTCCAGCTAGGAGCTGTCCGGCTAAGAGGAGTGGGCAGACGGGCTGCCT
GGAGCTGCCTCTGGCCGGTTGCGAGCCCGCCACCTGTCCAGGTGTGCCAGGGGTGGGGA
GCGGGAATCACTAGACCTCCGCCTCTGGCCCATCTGTCGTCGGCCTCTCCCGGCCCAGCTGCCCGGG
CCCGATCAAGGCGCCAGGGACTGGCCTAGGACCCCTTCCCGGCCCAGCGTCCCCCCTC
```

FIGURE 1 What type of map do we need to describe our genome? The map on the left shows the lie of the land around Ben Nevis—Scotland's highest mountain. Contour lines pick out precipitous cliffs and gentle slopes. The composition of different parts of the environment—rock, water, forest—are indicated through the use of symbols and colors. As well as a list of place names, this map also includes key structural and functional landmarks such as parking lots and rescue posts that are key to the way this landscape functions. The map on the right describes the DNA sequence of a part (0.000075 percent) of the human genome located on chromosome number 11. This sequence is part of the coding region for a gene called PAX6. Adapted from Bridger and Bickmore (1998, Figure 1).

were not uniform. In some places they were relatively frequent (every few hundred thousand base pairs apart), whereas in other parts of the map they were separated by millions of base pairs of DNA. This was not a random pattern—the areas where archipelagos of clustered islands were occurring with high frequency were at particular recognizable regions (bands) of chromosomes.

Chromosomes (from the Greek words meaning colored body) were first identified in the nineteenth century with the light microscope after staining with colored dyes. It is apt that the use of color permeates the way in which we analyze the chromosome and its structure today. In the early 1970s it was noticed that some stains produced reproducible bar codes on human chromosomes. These banding patterns enabled each chromosome to be unambiguously identified, and structural abnormalities in chromosomes could then be linked with particular diseases. This has been critical in our understanding of many genetic disorders. However, more fundamentally, these patterns suggest that there is an inherent organization of the genome at a level that affects both the structure and the behavior of the chromosome. We can crudely categorize chromosome bands into two types, "R" and "G" bands. From our elementary map it seemed that genes were clustering into the R type of chromosome band. This might have been a peculiarity of the small part of the genome that we were focused on at the time, rather than a generality. Therefore, we set out to discover whether this observation was true of the whole human genome: Are most of our genes concentrated into the parts of chromosomes that are visible to us down the microscope as R bands, and, conversely, are they thin on the ground in the G bands?

To answer this question we turned to a procedure known as fluorescence in situ hybridization (FISH). In FISH the chromosomal location of a particular DNA sequence is visualized by tagging it with molecules that are excited by, and emit, light of different wavelengths and hence different colors. The labeled DNA is then allowed to find its cognate site on chromosomes that have been spread out on glass slides, and the location of the colored "hybridization signal" is determined relative to the chromosome banding pattern using a fluorescence microscope. The labeled DNA molecules that can be used as probes in FISH can be chosen to light up not only single points on the chromosome, but also large regions of the chromosome, or even whole chromosomes. Probes that produce such broad brush strokes have appropriately been called chromosome "paints."

To find out where along the chromosome genes with CpG islands were to be found, we isolated all the islands from the human genome and labeled them en masse so that they would appear in red (Figure 2).

The pattern of red paint we saw by annealing human CpG islands to chromosomes dramatically showed us that genes are not uniformly dis-

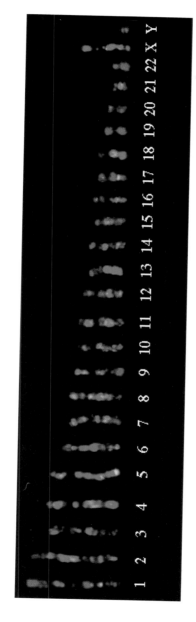

FIGURE 2   FISHing for a look at the organization of the human genome. The human chromosomes are laid out in order (karyotype), and the position of genes is painted in red. The green paint is used to show where the blocks of late replicating DNA are.

tributed either along, or between, our chromosomes. The areas of red paint did relate to the chromosome banding patterns; and not only did the density of genes, as measured by the intensity of the red paint, vary dramatically along the length of a single chromosome, but it also varied between different human chromosomes. Most striking were the differences between two small chromosomes—chromosomes 18 and 19. These are estimated to have similar amounts of DNA in them—approximately 70 million base pairs apiece. They also appear physically to be of the same size (human chromosomes are numbered in order of decreasing size). In our experiment, chromosome 19 was painted bright red whereas there was a distinct lack of red paint on chromosome 18. Our FISH experiment, which was designed to illuminate the gross organization of sequence within the genome, is being substantiated by the Human Genome Project itself as more and more individual genes (with or without associated CpG islands) are assigned to human chromosomes. In the databases at the moment, there are more genes localized to the small chromosome 19 than to any other chromosome in the karyotype barring the largest—chromosome 1. On the other hand, chromosome 18 has the smallest number of genes assigned to it of any chromosome except for the small Y sex chromosome. Our findings also allowed us to better understand why, excluding anomalies of the X and Y sex chromosomes, it is that only three human trisomies (additional copies of chromosomes 13, 18, or 21) commonly survive to birth. These are not the three physically smallest (in bases) chromosomes, but they are three chromosomes that our experiments showed as being poorly populated with CpG islands and that have small numbers of genes assigned to them. Trisomies of small but CpG island and gene-rich autosomes, such as chromosomes 19 and 22, are not generally seen as liveborne, and we assume that this is because their high gene concentration makes the presence of an extra copy incompatible with embryonic and fetal development and survival.

Our map of the human genome had already taken on new perspective: We could view the genome as broadly two different environments—one that is rich in genes and the other a gene desert. In some cases these two environments could be found intermingled within the same chromosome. In more extreme cases whole chromosomes seemed to be composed almost entirely of one or other type of environment. Does this represent a unique way of organizing a large and complex genome, or is it something that is found in other organisms and that might indicate that there are functional constraints on where genes can be placed on chromosomes? Studies of CpG island distributions in other vertebrates suggest the latter. The same high concentration of CpG islands into a limited number of distinct chromosomal locations is observed in the genomes of chickens, mice, and pigs. There appears to be a very good reason why genes prefer to hang around together in large groups.

COPYING THE MAP: WHERE TO START AND WHEN TO BEGIN

Before every cell divides to form two new daughter cells, the enzymatic machinery of the cell nucleus accomplishes a remarkable feat. Every single base of DNA in the genome is duplicated with precious few mistakes. This replication of the DNA takes place during a period of the cell cycle known as the S (synthesis) phase, which lasts for approximately eight hours in human cells. All of the DNA on the chromosomes must be copied accurately, but must be copied once only. The process of DNA replication must therefore be carefully controlled. Any mistakes—base changes, omissions, or repetitions—that are made, and that are not corrected, will be passed on to daughter cells as mutations in the DNA. The consequences of this can be genetic disease, genome instability, or tumor formation.

DNA replication begins from sites on chromosomes termed replication origins. If the replication of an average human chromosome were to start from a single origin in the center of the chromosome, it would take 17 days and nights to be completed. Clearly, more than one replication origin per chromosome is needed. An average chromosome may use about 1,000 origins to ensure that replication is finished within the duration of the S phase. At present, we have no way of picking out where the replication origins are by simply looking at the DNA sequence—and I believe that we may never be able to. Considerable evidence points toward the position of replication origins being encoded within the structure and shape of the chromosome, rather than in its sequence per se. Therefore, we will have to look to this level of genetic information in order to describe the position of replication origins on our map.

Not all replication origins fire off at the same time. DNA replication is under temporal control; some parts of the chromosome are consistently copied earlier than others. Evidence for this could be seen in our CpG island FISHing experiment (Figure 2) where the green banding pattern was generated by labeling the DNA that was copied during the last half of the S phase (late-replicating DNA). The red and green colors on these chromosomes do not overlie each other (we would see yellow if they did). Parts of the chromosome that have the highest concentrations of genes (red-painted regions) have their DNA copied in the early stages of the S phase (they do not take up any green color). Conversely, the DNA that is copied later (green) does not contain many genes (it is not red). For our two contrasting human chromosomes (18 and 19), the former is green, the latter is red. The same relationship between the presence of a lot of genes and the time of DNA replication is conserved in other vertebrates. Why might this be so?

During the S phase it is not only the DNA sequence that must be copied, but also the chromatin structure. The mechanism for making an accurate copy of the DNA molecule was readily apparent within the elegant structure of the DNA double helix itself. The same cannot be said for the replication of chromatin structures. We do not begin to understand how the different types of packing of DNA with protein, at different sites within the genome, can be copied and inherited (as epigenetic information) from one cell to the next, but we know that process takes place and that it is vital for the correct development of the animal. It is plausible that the time at which the DNA, and its associated chromatin, is copied may play some role in this. The spectrum of protein components that may be available to be chosen from to form new chromatin may be quite different at the early stages of the S phase than at the later stages. Hence, the very time of DNA replication could influence chromatin structure. This would ensure that the gene-rich compartments of the genome would become packed into a different sort of chromatin than the gene-poor (late-replicating) bits, compounding the differences in the environment of these two compartments of the genome. Our comprehensive map of the genome should include an element to describe the time of replication of different parts of the sequence.

GETTING ALL WRAPPED UP

The organization of the DNA sequence into chromatin is hierarchical: One layer of packing is superimposed upon another (Figure 3). What this compaction brings about is the shortening in length of the naked helix by a factor of 10,000-fold. At the foundation of this hierarchy is the wrapping of the DNA sequence, in short lengths, around proteins called histones. This forms a series of beadlike structures (the nucleosomes). The beads on a string are then further folded to form the basic chromatin fiber, with a diameter of three-millionths of a centimeter (30 nm) and a compaction of 100-fold in length over the starting DNA molecule. Histones are some of the most conserved proteins in evolution, reflecting their central role in underpinning all subsequent layers of chromosome packing. Despite this conservation of their basic sequence, there is a lot of variation built into the histones by chemical modification. So far, the modification that has received most attention is the addition of acetyl (CH_3COO) groups—histone acetylation. Addition of this seemingly innocuous chemical group to the ends of the histones has a dramatic effect on the structure of the result-

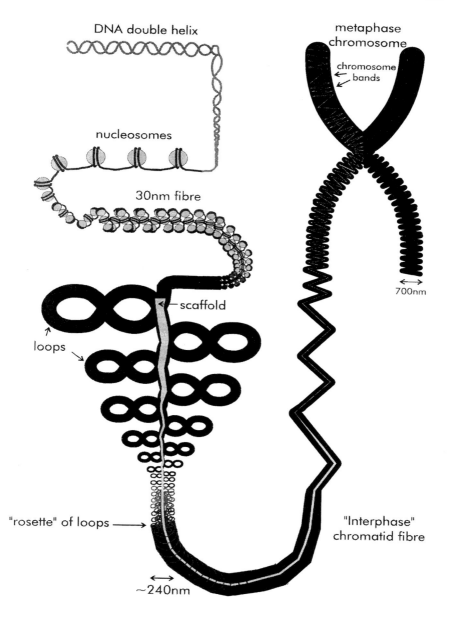

FIGURE 3 Folding up the map. The DNA molecule is packaged up with layers of protein like Russian dolls to form the final chromosome.

ant nucleosomes; the DNA associated with acetylated nucleosomes is much more accessible to other proteins and enzymes than is unacetylated chromatin.

In an experiment analogous to FISH, but using tagged antibodies to specific proteins as probes rather than DNA sequences (immunofluorescence), it was found that the most acetylated forms of chromatin in the human genome correspond with the R-band regions (i.e., the gene-rich, early-replicating compartment of our map). So, domains where genes are located are wrapped into the type of chromatin that is potentially most accessible to enzymes that are needed to transcribe gene sequences into the messages that eventually make proteins. The large parts of the genome with fewer genes are packed into more impenetrable chromatin. Predictably, human chromosomes 18 and 19 appeared unacetylated and highly acetylated, respectively, in this experiment.

Why bother to make this distinction in chromatin? Why not just package the whole genome into accessible acetylated nucleosomes? First, it may help sequester the gene-barren DNA away from proteins whose job it is to find genes and so limit their search area. It also opens up the opportunity to put a few special genes, whose inappropriate expression might be detrimental, into an environment that ensures their silence until the appropriate moment. It also adds another layer of control to gene expression that can operate over large chromosome areas rather than on a gene-by-gene basis. There is one particularly striking example of this: The silencing (inactivation) of genes on one of the two X chromosomes of females. This process is thought to ensure that there are equal doses of gene products from X-linked genes in males (who only have one X chromosome) and females (who have two). Once the X inactivation process has begun, early in the development of female mammals, the first detectable change in the chromosome is that its time of replication becomes late. Then the chromatin loses its histone acetylation, and many, but not all, of the genes on the chromosome fall silent. This might be a good example of a case in which a change in replication timing forces an alteration in the protein composition of the chromatin. Other late-replicating regions of the genome, that also lack acetylation, are completely gene-free areas where the DNA consists of simple short sequences that are reiterated over and over again. Some of these sites, though not coding for genes, may have very important roles in maintaining the structure and integrity of chromosomes. In the conventional human genome map, these regions will look very monotonous—the same sequence motif of a few hundreds of bases is reiterated again and again over many millions of bases—but in our version of the map their structure and topology will be important and distinctive components of the genome landscape. In particular, the centromere—the site on the chromosome that ensures the accurate segregation of the genome into daughter cells—is composed of this sort of repeating DNA and is also late replicating and is not acetylated. Its more compact chromatin structure may be critical in providing solid bedrock for a part of the chromosome that has to endure strong mechanical forces as part of its job. Each chromosome's worth of a double-stranded DNA molecule is called a chromatid, and by the time the cell has replicated the whole genome and is ready to divide, there are two sister chromatids per chromosome that are held together at the centromere. The centromere's functions are both to hold onto the two sisters until the right moment before letting go, and also to interact with the cagelike apparatus called the spindle that ensures each new daughter cell receives only one chromatid from each pair of sister chromatids—one accurate genome's worth of DNA. During cell division, chromosomes jostle in the middle of the spindle until one chromatid of a pair attaches to one side (pole) of the cage, whereas the other attaches to the opposite pole. Forces from the spindle poles pull on the chromatids, like a perfectly balanced tug of war. The all-clear is then given for the sister chromatid attachments at the centromere to be broken and for the chromatids to be pulled by motor proteins toward opposite poles of the spindle. If any part of this process—the making or breaking of attachments—ever goes awry, the consequences are dire. Cells in our bodies with too many or too few chromosomes can progress to form tumors. If the same thing happens in the formation of sperm or egg, the resulting fetus—if it develops at all—may carry a trisomy and suffer severe defects.

There are specific enzymes whose job it is to put acetyl groups onto histones—the acetyl transferases—and those whose job it is to undo this—the histone deacetylases. The overall level of acetylation in chromatin is the playoff between these two opposing activities. If this balance were altered between early and late S phase, then it is easy to see how the time of replication might influence the histone acetylation level of a particular part of the genome. Nor does this train of events seem to be all one way: The histone acetylation level of chromatin may have some influence on when a piece of the genome is replicated. We can experimentally interfere with the normal balance of histone acetylation using chemicals that specifically inhibit histone deacetylases. We had been studying the replication behavior of parts of the human genome that are said to be imprinted. That is, the decision as to whether genes in these chromosome regions will be expressed depends on which parent the chromosome was inherited from. Some genes are only ever expressed when they are inherited from fathers, and for other genes it is the maternally derived copy that is active. This is not due to any difference in the DNA sequence per se between genes inherited from either parent: The conventional map of the human genome could look the same for the genome you inherited from your mother and for the one that you got from your father, but the readout from each map will be very different. It is important that you do inherit one chromosome copy and its genes from a sperm and one from an egg. If you do not, and through a mistake in the partitioning of chromosomes you happen to get both of your copies of chromosome number 11 from your father, you will develop abnormally. Parts of your body will be overgrown and you will be at risk of developing certain cancers. How do the chromosomes in the cells of our bodies still remember which parent they originally came from so many cell divisions ago? Because we know that the memory is not in the DNA sequence, it must be imposed epigenetically. Some of the memory lies in the chemical modification (by methylation) of C residues in the DNA sequence. The pattern of methylation can be accurately copied during DNA replication. However, we also provided evidence that part of this memory is in the histone acetylation state of chromatin. We observed that imprinted parts of chromosomes are also replicated at different times, depending on the parental origin. We calculated this difference in time to be less than one hour—a much more subtle difference in timing than that seen between the gene-rich and the gene-poor domains along chromosomes (several hours). We found that we could wipe away this memory if we stopped the histone deacetylases from working (the parental copy that had been the later replicating one caught up with the earlier replicating one). Most excitingly, in collaboration with two groups in Cambridge, United Kingdom, we found the same pattern of synchronous replication (but in the absence of any experimental interference with acetylation), at regions that should have been imprinted, within the genomes of individuals who had all the clinical symptoms of an imprinting defect. These individuals looked as though they had got two copies of chromosome 11 from their fathers by mistake, but they had not. They had inherited one paternal and one maternal copy of chromosome 11. What must have been wrong in these individuals is that the maternal copy of the chromosome had forgotten where it had come from and so incorrectly expressed genes that should normally only be read from a paternal chromosome 11. The methylation pattern of the chromosomes looked normal. Could the underlying imprinting defect in these individuals be a problem with the pattern of histone acetylation?

FOLDING LANDSCAPES

The interaction of DNA with histones is far from the end of the story as far as the chromosome is concerned. A further tenfold to a thousandfold of packing still has to take place (Figure 3), and how this is achieved is a matter of contention. A growing consensus is that the chromatin fiber is organized into a series of loops, which radiate out from a group of proteins collectively known as the chromosome scaffold (Figures 3). The loops are probably wound up on themselves, like twisted rubber bands, in the final fully condensed form of the chromosome. Use of the term scaffold immediately conjures up an image of a structural element providing a framework for the underlying form, and indeed the chromosome scaffold may serve no purpose other than to aid in the spatial management of the genome and to add rigidity to the chromosome. However, I find it an appealing idea that the topological constraint of DNA into loops, anchored at their bases, should naturally organize the genome into functional units. What might these units be?

The internal architecture of the chromosome can be revealed experimentally by biochemical extraction of proteins from the chromosome. This reveals the residual scaffold of proteins at the core of the chromosome and running along the length of the chromosomes. The main proteins in this structure are enzymes that are capable of either disentangling strands of DNA or of coiling, folding, and condensing chromatin—all sensible activities for a central element involved in organizing the chromosome and in separating its sister chromatids ready for cell division. In these partially unraveled chromosomes, the loops of stripped chromatin spread out from the scaffold to a far greater extent than can be seen in native chromosomes. We reasoned that if such a deconstructed chromosome could still be used in FISH experiments, then, because we had preserved some of the two-dimensional structure of the chromosome, we would be able for the first time to see some of the topology of the DNA sequence within the chromosome. This was indeed the case: The winding path of long stretches of the human genome could be traced using tags of different colors (Figure 4). Specific places along the DNA sequence could be mapped as being close to the scaffold at the center of the chromosome. Other sequences were exposed on the chromosomes's surface. At the chromosome scaffold we saw that the DNA sequence is bent sharply so that these sites are in the bottom of steep-sided valleys of chromatin, and the sequences at the apex of the loops are on the hilltops of the genome.

Like a reflection in a lake, the topology of one DNA molecule was mirrored in its sister. This implies that the folding of the genome is copied just as the sequence is copied. How this can be brought about is beyond the limits of our current knowledge. What was most intriguing about our observation was that the places on the chromosome where we knew origins of replication were located were in the valley bottoms close to the chromosome scaffold. The frequency with which the hills and glens occurred in the chromosome appeared to be quite regular—the troughs were every few hundred thousand bases so that the rolling hills of the

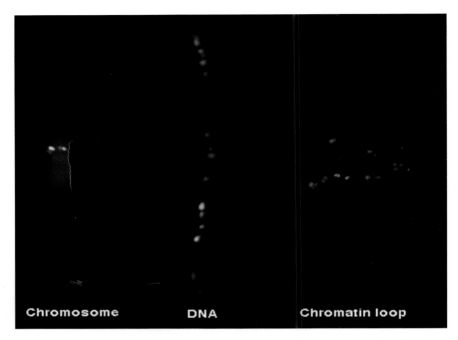

FIGURE 4   The first two-dimensional map of the human genome? The three images are from a FISH experiment in which the same two DNA probes were used. The first—500,000 bases in length—was labeled in red; the green probe is 40,000 bases long and is located within the sequence covered by the first probe. When these two probes are hybridized together to a fully condensed chromosome (blue), the DNA packing is so tight on both sister chromatids that the signals from the two probes are superimposed on each other in tight spots. At the other end of the spectrum, when the two probes are hybridized to naked DNA, from which all chromatin has been removed, the signal is a long line of red signals with the green signal part way along this line (central image). On the right, the probes have been hybridized to stripped chromosomes in which only the scaffold remains. The topology of this part of the genome can now be resolved. The part of the sequence associated with the scaffold is halfway along the red signal and is sharply bent. The green signal is halfway up the slope made from a chromatin loop. Adapted from Bridger and Bickmore (1998, Figure 1).

genome are all of about the same height. This frequency is surprisingly similar to that of replication origins, and other organisms in which the valley periodicity is different from that of our own genome have paralleled changes in replication origin frequency. We may have found a signature in the topology of the genome that identifies origins of replication.

There was one exception to this vista of rolling hills throughout the genome, and that was at the position of the centromere. Even though the simple DNA sequences at each human centromere extend over many millions of bases—enough to make several hills' worth of chromosome loops—we could not see any. This entire DNA was constrained tightly to the chromosome scaffold. Is this another example of structural rigidity being built into the centromere through an unusual chromatin structure? We also saw another unique topological feature at the centromere that appears to relate to centromere-specific functions. At all other sites in the genome that we looked at, the DNA molecules of the two sister chromatids were close together but were clearly physically separate. However, at the centromere we could see strands of DNA tangled up in between the two sisters. Remembering that one of the functions of the centromere is to hold sister chromatids together until the right time, we thought that this entangling of centromeric DNA might be part of the molecular glue holding the sisters together. There seems to be no other part of the genome in which the relationship between genome topology and function is so graphically displayed as at the centromere.

INTO THE THIRD DIMENSION

The map of the human genome described here is growing in complexity and is taking on elements of color and shape. However, it is still a map in a vacuum whereas the genome operates inside a membrane-bound cell nucleus—the central part of the cell in so many ways. Therefore, we need to describe not just how the chromosome looks with respect to its own components, but also the relationship of the genome to the nuclear space. Also, the nucleus itself is not a homogeneous bag. We are increasingly aware that the nucleus is organized into compartments, each with distinct functions. In retrospect, spatial management of the nucleus into compartments is probably the most sensible way to deal with a complex genome and its diverse functions, but it has taken us a long time to come to this realization. This is because, unlike the very obvious compartments in the rest of the cell (cytoplasm) each bounded by membranes, those in the nucleus do not have any very obvious morphological or physical boundaries.

Many textbooks have a lot to answer for in their description of the chromosome inside the nucleus—usually some variation of spaghetti on a plate is common. It is assumed that the DNA molecules of each chromosome are splayed out all over the available space and intertwined with each other. We now appreciate that each chromosome is a rather distinct body, occupying a discrete space or territory within the nucleus—in fact more penne than spaghetti! There is no evidence that any particular chromosome has a definitive address within the nucleus, but rather that there are preferred locations for certain chromosomes. Probably the best example is five of the human chromosomes (numbers 13, 14, 15, 21, and 22). These are all physically associated with each other because each of them houses copies of the same genes—the ribosomal RNA (rRNA) genes. The RNA made from these genes is an essential component of the cellular machinery that translates conventional RNA messages from genes into protein (i.e., it decodes the genetic code). RNA is processed for this job in the best understood of the compartments of the nucleus—the nucleolus (literally, the little nucleus)—and these five chromosomes congregate around it with their rRNA genes protruding into it. These five chromosomes very well illustrate the concept that DNA sequences that might be far apart in the conventional genome map—even on different DNA molecules—can be physically close in space. This physical closeness also manifests itself through plasticity of the genome in the form of chromosome translocations. Translocations are exchanges of large portions of sequence between different chromosomes so that, for example, part of a sequence from chromosome 1 appears on chromosome 2, and the displaced chromosome 2 material is now on chromosome 1. These sorts of large-scale movements have probably been a major driving force in genome evolution, and they are also a common cause of genetic disease because they involve breaking and rejoining the DNA. If the break occurs within a gene sequence it can prevent the production of the correct protein product from that gene. Alternatively, altered and inappropriate gene and protein function can be formed by the fusion of two gene sequences that were originally on different chromosomes. This underpins many translocations associated with human leukaemias. Given that exchanges of sequence between separate DNA molecules are involved in chromosome translocations, it is not unreasonable to assume that the frequency with which they occur is a reflection of physical proximity of the chromosomes in space (i.e., in the nucleus). In accord with this, the most common translocations in man are those occurring between the rRNA-carrying chromosomes. Can we use the frequencies of other chromosome translocations in the human population to help us build our three-dimensional map of the genome?

Because of their physically similar size but different functional characterics, we asked what the respective territories of chromosomes 18 and 19 looked like in the nucleus (Figure 5). They were surprisingly different. Chromosome 18 was considerably more compact than chromosome 19, despite their similar DNA content, and they also differed in their relative dispositions. Chromosome 18 was found toward the periphery of the nucleus, and chromosome 19 was located more centrally. Because most of the volume of a sphere is around its edge, chromosome 18 is in the part of the nucleus where the majority of the genome is located. Chromo-

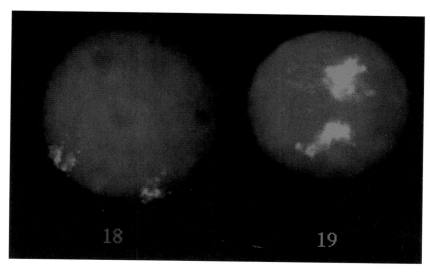

FIGURE 5   Despite the fact that human chromosomes 18 and 19 each contain about 70 million bases of DNA, the two chromosomes 19 (green) adopt much bigger territories within the nuclei of our cells than do the pair of 18s (red). Chromosomes 19 are also more centrally located in the nucleus and make more contacts with the nuclear matrix than do chromosomes 18.

some 19 is in a more confined location and so might be expected to come into proximity with other chromosomes less frequently. As expected, chromosome 18 is found as the partner in translocations to other chromosomes much more frequently than is chromosome 19. Rarest of all are translocations between chromosomes 18 and 19, but we were able to find one example. In nuclei of these cells, the translocation chromosomes took up distinctive orientations so that their chromosome 18-derived material was oriented toward the chromosome periphery whereas their chromosome 19 component pointed to the middle. Each part of the translocated chromosome seemed to know where its real home should be.

If, as we showed in our experiments on the chromosome scaffold, some parts of our DNA sequence are buried in the chromosome interior, whereas other parts are exposed on the surface, then it follows that these two sorts of sequence will see the nucleus and its components in quite different ways. Sequences on the outside will be in intimate contact with the nuclear environment. Sequences in the chromosome core may be quite protected. It has been argued that the various bits of machinery that are needed to express genes and to process their messages are embedded within a network of proteins in the nucleus called the matrix. This matrix would form channels running between the chromosome territories. Active genes are thought to make more contacts with this matrix and to be exposed on the chromosome surface. The RNA products of these genes can then be transported out of the nucleus and into the cytoplasm through the interchromosomal channels. In accord with this idea we have found that human chromosome 19 makes more intimate contacts with the nuclear matrix than does chromosome 18.

THE GEOLOGY AND GEOGRAPHY OF THE GENOME

To describe our map of the genome we need to understand what all the components of the complex landscape are. Surprisingly we are still a long way from this simple goal. Some of the most abundant nuclear and chromosomal proteins, and some of those with very specific and tractable functions, have been identified. We have recently begun a series of experiments to augment the components list for the chromosome and nucleus. This experiment takes advantage of the fact that we can see chromosomes, and even parts of chromosomes, with the aid of a microscope. We have randomly put into the mouse genome 9,000 bases of sequence that, when inserted appropriately into an endogenous mouse gene, will tag the end of the protein product of that gene with the ability to turn a simple chemical into a blue stain. This stain is deposited in the cell wherever the tagged protein normally resides. In this way we have already identified three new proteins that reside in different parts of mammalian chromosomes. The design of the experiment also allows us ready access to the sequence of the "trapped" host genes. Database analysis with these sequences has exposed our ignorance about the protein composition of chromosomes. The sequences of all three new genes encoding chromosomal proteins match human and mouse DNA sequences that have been recorded as being genes of a previously unknown function. We can now assign a partial function to these genes, based on the location of their protein products within the cell. There is a certain irony to this: We are using the current human genome map and sequence database, and that of the mouse, to describe more fully the human genome! This is a step toward the functional genomics phase of the human genome project. One of the fundamental things about our genome that we are finding out from this experiment is that mammals appear to devote a larger proportion of their genes to the job of making nuclear proteins than do simpler organisms such as yeasts. This suggests to us that much of the complexity of multicellular animals has been brought about not so much by expanding the repertoire of different proteins in the cytoplasm or at the cell surface, but by making the control of gene expression in the nucleus in time and space more exquisite.

MOVING MAPS INTO THE FUTURE

Maps are not very good at depicting the fourth dimension, but we do need to build an element of time into our maps of the future. The genome moves and changes shape through the cell cycle and as cells divide. We have been hampered by the fact that most of the ways that we have looked at the genome depend on killing the very cells we are studying. At last we now have a way of potentially following different parts of the genome in real time and space in living cells. This advance came from the unlikeliest of sources—a jellyfish. A protein found in this animal is naturally fluorescent—it shines out in green when exposed to ultraviolet light. The sequence of the gene encoding this protein was cloned and adapted for use inside mammalian cells (Figure 6). We have begun experiments

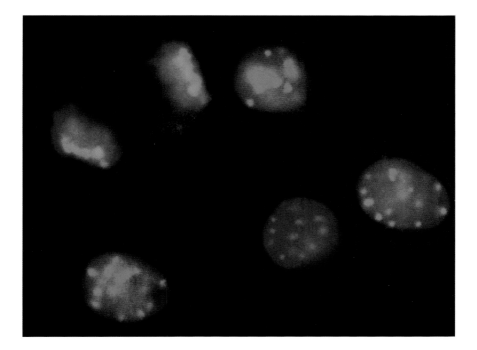

FIGURE 6 Living mouse cells in which the centromeric regions of the genome are lit up by green fluorescent protein (green) and the nuclei and chromosomes are stained in blue. In the top left is a cell that is in the process of dividing. The two chromosome sets are migrating toward opposite poles of the spindle (which we cannot see) with the green centromeres clearly leading the way and the rest of the genome following.

targeting green fluorescent protein to different parts of the human and mouse genome. With it we look forward to making bright, multicolored, and multidimensional maps of the human genome in the future.

SUMMARY AND CONCLUSIONS

I have described our vision of the human genome as a complex, heterogeneous, highly organized, and folded but moving environment. Is this just an aesthetic picture of the genome or of an important substance for us to understand? Normal gene expression and development depend on the maintenance of correct nuclear and chromosomal organization. In some instances, this organization must be robust enough to persist through many cell divisions, but in other cases it must be flexible and responsive to signals from both within and outside of the cell. Moving a gene out of its normal context and placing it somewhere else in the genome with a different environment can result in the failure of that gene to work correctly—a so-called position effect. An increasing number of genetic disorders are being shown to result from this type of positional defect, rather than from a mistake in gene sequence. The key to understanding them will lie in chromosome and nuclear structure. Similarly, I have illustrated how key chromosomal functions, such as the ability to replicate and segregate the genome depend on the folding of the DNA sequence. The very integrity of our genome depends on the successful execution of these functions throughout our lifetime—we each produce one hundred thousand million new blood cells and cells to line our gut every day. Defects in the ability to segregate chromosomes accurately through these many divisions and to maintain the integrity of the genome underpin many human cancers, infertility, and developmental abnormalities. There is hope that in the near future many genetic defects in man can be corrected by gene therapy. Normal copies of defective genes will be put into cells on small artificial chromosomes. For this to work to its full potential we need to understand how normal genes are organized and controlled so that we can produce appropriately functioning engineered versions for any particular cell or tissue type. If we better understand the diverse environments of genes in their normal home, we might also be able to tailor artificial chromosomes to have similarly different surroundings and to adopt different positions in the nucleus to suit the needs of different applications.

Determining the sequence of the human genome will be the end of a massive undertaking, but it is only the beginning of a long and winding road in efforts to understand how we evolved, how we develop, and how we maintain and pass on our genomes.

REFERENCES

Bridger, J. M., and W. A. Bickmore. 1998. Putting the genome on the map. *Trends in Genetics* 14:403–409.

Paulson, J. R., and U. K. Laemmli. 1977. The structure of histone-depleted metaphase chromosomes. *Cell* 12:817–828.

9

Human Genetics in the Twenty-First Century

Leonid Kruglyak

ABSTRACT

As human genetics enters the twenty-first century, many powerful molecular tools are either in place or about to arrive. They include a large capacity for DNA sequencing, an ability to examine many polymorphisms in many people, and techniques for monitoring expression levels of many genes under many conditions. These tools are already generating large amounts of data. Early in this century, we can envision obtaining complete DNA sequences of many individuals, expression levels of all genes in a variety of cells, and a complete characterization of protein levels and interactions. The key task facing human genetics is interpreting this data.

My research centers on addressing different aspects of this task. These include understanding the genetic basis of diseases and other phenotypes, the evolutionary history of human populations, and the regulatory networks of gene expression.

Genetic Basis of Disease

Genetic factors are thought to play some role in susceptibility to almost all diseases. We can search for these factors using several approaches. The most well-known entails tracing inheritance patterns of disease in families in order to implicate particular chromosomal regions

Division of Human Biology, Fred Hutchinson Cancer Research Center, Seattle, Washington

in disease transmission. Such studies have produced many important advances in our understanding of the genetic basis of diseases and other phenotypes, particularly those with simple Mendelian inheritance. For more common diseases, such as diabetes, heart disease, cancer, multiple sclerosis, asthma, manic depression, and schizophrenia, successes have been harder to come by because of the problems posed by genetic complexity. The genetic basis of such diseases is likely to consist of a large number of subtle effects, whose detection may require a prohibitively large number of families. Population studies offer a potential way to overcome this hurdle. The most direct approach is to start with a mutation and examine its effect on frequency of disease in populations. Alternatively, we can recognize that all people share ancestry sufficiently far back in time, and that many mutations in today's population are likely to have a common origin. We can therefore search the DNA of population members for ancestral signatures of this origin.

Population History

The best repository of historical information is the DNA of population members. Patterns of spelling changes in the genetic code contain a record of population divergences, migrations, bottlenecks, replacements, and expansions. In addition to its inherent interest, this information provides important insights into the use of populations for studies of the genetic basis of disease, including the specific advantages offered by different populations. Molecular approaches have already contributed a great deal to our understanding of human evolution. These approaches need to be updated to take full advantage of modern molecular tools, provide a more detailed look at the entire genome, and resolve finer points of population diversity and relationships.

Regulatory Networks

The genetic code is a static record of information. The dynamic response of a cell to changing conditions is governed in large part by which genes are expressed—transcribed into messenger RNA. Several emerging technologies now allow simultaneous measurements of transcription levels of many genes in a cell under a variety of conditions. A key task is to reconstruct the network of interactions among genes from these observations. Understanding such regulatory networks is an important part of studying any biological process. Going beyond gene expression, many important dynamic responses involve modifications and interactions of proteins. Once technology allows proteins to be monitored with ease, ideas developed to understand networks of interacting genes can be applied to this next level of complexity.

Genetic Research Must Be Applied Responsibly

A key aspect of this goal is education. We must make our research and its implications clearly and broadly understood. Armed with appropriate information, society will be prepared to face the complex questions posed by progress in human genetics, reaping the benefits while avoiding the pitfalls.

Genetic approaches extend beyond human genetics to studies of pathogens and organisms of agricultural importance, and even more broadly to a better understanding of genetics and biology of all organisms. The ultimate benefits of genetic research to society take many forms: better medicine, improved quality of life, increased knowledge, and a greater understanding of who we are and where we fit in among other forms of life.

INTRODUCTION

The second half of the twentieth century witnessed the development of molecular biology and its tremendous impact on our understanding of life. In the past two decades, molecular techniques have been applied to our own species, giving rise to human molecular genetics and the resulting revolution in medicine. The Human Genome Project is poised to generate a complete readout of the human genetic code by the year 2003. It is not too early to ask: How much of the secret of human life is contained in the complete DNA sequence of every person on the planet, and how can we extract it? My research centers on answering this question in its many forms, a task that will occupy human geneticists for much of the twenty-first century.

This scenario of obtaining the complete DNA sequence of every person on the planet is not entirely science fiction. Already, new technologies are allowing significant fractions of the genetic code to be re-read and checked for spelling differences in large numbers of individuals. It is not far-fetched to imagine that early in the next century this will become possible for the entire genome—although probably not with the ease portrayed in the recent film *GATTACA*. Even today, the thought experiment of what we can learn from this information places an upper bound on what can be done with more limited data.

The central question posed above has many facets. One concerns the genetic basis of disease. Genetic factors are thought to play some role in susceptibility to almost all diseases. Which spelling differences (mutations) in the genetic code cause or increase susceptibility to which diseases? We can attack this problem using several approaches. Perhaps the most well-known entails tracing inheritance patterns of disease in families in order to implicate particular chromosomal regions in disease transmission. These regions can then be scoured for the causative mutations. Another approach is to start with a mutation and examine its effect on frequency of disease in populations. Alternatively, we can recognize that all people share ancestry sufficiently far back in time, and that many mutations in today's population are likely to have a common origin. We can therefore search the DNA of population members for ancestral signatures of this origin.

These approaches can be extended from studies of disease to those of normal human variation. They can also be applied to other species, including plants and animals of agricultural importance, model organisms of medical or biological interest, and pathogens.

Another facet concerns the evolutionary history of human populations. It appears likely that all of today's populations descend from a single, small ancestral population that lived on the plains of East Africa roughly 100,000 years ago. The pattern of spelling differences in our DNA contains signatures of subsequent population divergences, migrations, bottlenecks, replacements, and expansions. If we can understand how to read these signatures, we can answer many interesting questions about our history, as well as about the genetic diversity of and the relationships among today's populations. In addition to their inherent interest, the answers provide important insights into the use of populations for studies of the genetic basis of disease, including the specific advantages offered by different populations.

The genetic code is a static record of information. It alone does not tell us about the dynamic circuitry of a cell. The response of a cell to changing conditions is governed in large part by which genes are expressed—transcribed into messenger RNA. Understanding the regulatory networks of gene expression is thus a key part of studying any important biological process.

A final facet concerns our responsibilities in dealing with the consequences of genetic research. The key responsibility is education—we must make our research and its implications clearly and broadly understood. Armed with appropriate information, society will be better prepared to face the complex questions posed by progress in human genetics, reaping the benefits while avoiding the pitfalls.

With this survey of the landscape, I now examine the different facets in turn.

GENETIC BASIS OF DISEASE IN FAMILIES

Many diseases run in families. The list includes both diseases with clear genetic causes and simple patterns of inheritance, such as cystic fibrosis and Huntington's disease, and diseases with more subtle genetic predispositions, such as heart disease and schizophrenia. To date, most of the progress in our understanding of the genetic basis of disease has come from tracing inheritance patterns in families in order to implicate particular chromosomal regions, and ultimately to identify the culprit genes.

Linkage Studies: The Principle

Family studies rely on the principle of genetic linkage, first elucidated by Sturtevant and Morgan in their work on the fruit fly *Drosophila* in the early part of this century. Suppose a disease gene is segregating in a family, with all affected members inheriting a mutated copy from a single ancestor. As the chromosome on which the mutation resides is transmitted from one generation to the next, regions close to the mutation tend to be inherited with it, while those further away frequently recombine, as illustrated in Figure 1. A correlation is thereby established between genetic markers near the mutation and the mutation itself. One can therefore determine the general location of the disease-causing gene by looking for markers whose transmission correlates with inheritance of disease. Once a gene is thus localized, its position can be further refined by a range of techniques, leading eventually to identification of the gene itself and the disease-causing mutations.

Linkage Studies: Genetic Markers

In *Drosophila*, simple Mendelian characteristics such as eye color serve as genetic markers. A catalog of many such characteristics allows straightforward genetic localization of new phenotypes using crosses. This approach does not work in humans because we cannot set up crosses, and there are few common polymorphic characteristics (blood type and immunocompatibility loci providing some important exceptions). For many years, this lack of genetic markers hampered the usefulness of linkage for mapping human disease genes—the theoretical principle was waiting for the molecular tools to catch up. The situation changed dramatically with the proposal in 1980 by David Botstein and colleagues that spelling changes in DNA, which are highly abundant, can be used as markers (Botstein et al., 1980). Genetic maps covering all the chromosomes were

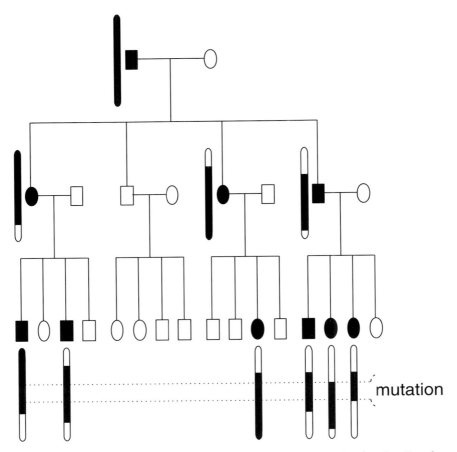

FIGURE 1   Transmission of a dominant disease gene through a family. For clarity, only the mutation-bearing chromosomes in affected individuals are shown. Shaded segments represent pieces of the chromosome on which the mutation entered the family. Dotted lines delimit the region to which the mutation can be confined based on the information in this family.

rapidly developed, and we have seen the ensuing progress in mapping and cloning Mendelian disease genes. Genes for over 700 diseases have been localized to chromosomal regions, and over 100 of these have been identified, including genes involved in cystic fibrosis, Huntington's disease, muscular dystrophy, and hemochromatosis.

Common Diseases: Genetic Complexity

Mendelian diseases are rare. Even the most common ones such as cystic fibrosis affect no more than one in a thousand people. Many others occur only in a handful of families. In total, Mendelian diseases are thought to affect roughly 1 percent of the population. By contrast, diabetes alone affects about 2 percent of people worldwide. Diabetes has a strong genetic component: A child of a diabetic faces a fivefold increased risk of developing diabetes. Genetic factors also play an important role in other common diseases, such as heart disease, cancer, multiple sclerosis, asthma, manic depression, and schizophrenia. Clearly, genetics can have its biggest impact on medicine by improving our understanding of common disease causation. Yet progress in this area has been modest compared with that in the study of Mendelian diseases. Why?

In the case of common diseases, the linkage approach runs into the problem of genetic complexity. In simple Mendelian diseases, mutations in a single gene always lead to disease and are responsible for all cases of the disease. In the case of genetically complex phenotypes, which encompass virtually all common diseases, this perfect correlation between genotype and phenotype breaks down in one or more of several ways. First, mutations in a gene may lead to disease only some of the time, depending on environmental, other genetic, or random factors. Second, some disease cases may reflect nongenetic causes. Third, mutations in any one of several genes may cause the same disease, or mutations in multiple genes may be required for disease causation. As a result of these and other problems, linkage studies of common diseases are much more difficult than those of simple Mendelian disorders.

One approach to overcoming the problems posed by complexity is to look for families in which a common disease approximately shows Mendelian single-gene inheritance. A prominent example of this approach is the successful identification of breast and ovarian cancer genes BRCA1 and BRCA2. However, this approach may not tell us about the genetic basis of a disease in the general population. Also, families with simple inheritance have not been found for many important diseases.

Common Diseases: Interpretation of Results

Despite the difficulties outlined above, recent years have seen frequent announcements that a gene "for" schizophrenia or alcoholism or reading disability has been found. Many of these findings have not stood the tests of time and replication, whereas others remain a subject of contention. Why has it been so difficult to establish whether a correlation between a chromosomal region and a common disease is real or a statistical artifact? In simple Mendelian disorders, we know that a single gene accounts for all cases of disease. In the case of common diseases, the number of susceptibility genes and the magnitude of their effects are unknown prior to analysis. A typical study employs a very large number of statistical tests in screening the entire genome for a possible effect. Smaller effects are obscured by random fluctuations in inheritance. As a result, the "signal" is easily confused with "noise."

To ease the problem of interpretation, Eric Lander and I developed guidelines for reporting results from family studies (Lander and Kruglyak, 1995). The guidelines attempt to strike a balance between avoiding too many false results and not missing true effects. A critical application of the guidelines allows us to provide a realistic picture (albeit with blurry edges) of what we really know about the genetic basis of common diseases. New reports on CNN that a gene for disease x has been identified can be treated with appropriate caution.

Common Diseases:
Unknown Mode of Inheritance and Small Effects

A simple Mendelian disorder shows recessive, dominant, or sex-linked inheritance. By contrast, the mode of inheritance of common diseases is typically unknown. An incorrect guess about the mode of inheritance can cause a gene to be missed. Genes involved in common diseases also have relatively small effects on overall disease susceptibility. Detection of such effects requires studies with much larger sample size, placing greater importance on efficient analysis of the data. Over the past several years, a large part of my research effort has focused on providing human geneticists with more powerful tools for detecting complex genetic effects (Kruglyak and Lander, 1995, 1998; Kruglyak et al., 1995, 1996). I have developed statistical techniques that can detect susceptibility genes without having to make guesses about the unknown mode of inheritance. I have also developed computer algorithms and software that allow much more efficient and rapid analysis of family studies. Unlike earlier methods that focused on single genetic markers, the new methods are explicitly designed to take full advantage of the greater power of today's dense genetic maps. My computer program GENEHUNTER has been used to find chromosomal regions that may be involved in a number of diseases, including diabetes, schizophrenia, inflammatory bowel disease, multiple sclerosis, and prostate cancer.

Interlude: Genetics of Mosquito Resistance to the Malaria Parasite

Linkage approaches can also be used in nonhuman organisms to study a number of important problems. Malaria affects hundreds of millions of people worldwide, primarily in developing countries. The parasites that cause malaria are transmitted to humans by mosquitoes, within which they must spend a stage of their life cycle. Thus, one approach to combating malaria is to interrupt the chain of transmission by making mosquitoes resistant to the parasites. Some mosquitoes are naturally resistant, whereas others are susceptible, with genetic factors playing a clear role in this variation.

Together with Ken Vernick, Yeya Toure, and Fotis Kafatos, we have launched a project to study the genetic basis of parasite resistance in the mosquito population of Mali. In the study, wild female mosquitoes are caught in a local village and allowed to lay eggs in captivity. Lab-reared progeny are then infected with natural populations of the malaria parasite and examined for resistance or susceptibility. Once this phase is completed, we will search for linkage between genetic markers and parasite resistance in mosquito families. The study is similar to family human studies in that mosquitoes are drawn from an outbred population, rather than from inbred laboratory strains. This design allows detection of genetic factors that are important in the wild. An important advantage of the mosquito system, which should help overcome the genetic complexity of parasite resistance, is the ease of generating very large families—a single female mosquito can produce over 100 progeny. The results of the study will contribute to a better understanding of the mosquito–parasite interaction and may ultimately permit the development of a genetic strategy to interfere with malaria transmission.

Summary

Family studies have produced many important advances in our understanding of the genetic basis of diseases and other phenotypes, particularly those with simple Mendelian inheritance. Although some tantalizing hints regarding the genes involved in more common disorders have emerged from family studies, successes have been harder to come by because of the problems posed by genetic complexity. It is now becoming increasingly clear that the genetic basis of common diseases is likely to consist of a large number of subtle effects, which interact with each other and the environment. Detection of such effects by linkage requires a prohibitively large number of families. Population studies may offer a way to overcome this hurdle.

GENETIC BASIS OF DISEASE IN POPULATIONS: COMMON VARIANTS

Example: Alzheimer's Disease

Alzheimer's disease is a devastating neurological disorder that strikes about 1 percent of people aged 65–70 and up to 10 percent of people over 80 years of age. Its prevalence is on the rise as the population ages. Progress in understanding and treatment of Alzheimer's disease is thus a key step in extending healthy life span. Family studies have identified several genes that cause early-onset Alzheimer's disease in some families, but these genes contribute to only a small fraction of all cases. What are the genetic components of the much more common late-onset disease?

Several lines of evidence led to the Apolipoprotein E (APOE) gene. This gene has three common variants, or alleles: E2, E3, E4, each one differing by a single nucleotide. Studies of the APOE gene in Alzheimer's patients and healthy controls showed that although the control frequency of the E4 allele is about 15 percent, it goes up to 40 percent in patients. Each copy of the E4 allele raises a person's risk of Alzheimer's disease roughly fourfold. Variation at the APOE locus appears to account for about 50 percent of the total contribution of genes to Alzheimer's disease. Interestingly, variation in the APOE gene also contributes to cholesterol levels and heart disease susceptibility, and may play a role in longevity.

Association Studies: Basic Concept and Advantages

The APOE–Alzheimer's story is an illustration of the simplest type of population studies, known as association studies. These studies directly examine the effect of a particular variant on a disease (or other phenotype). In the future, such studies will be conducted by resequencing entire genomes and looking for consistent differences between patients and controls. Until this becomes feasible, the strategy is to find common variants in genes and examine their phenotypic effects in populations. The approach therefore relies on the existence of such common variants, as well as on our ability to detect them.

Association studies have important advantages over family linkage studies. Instead of searching for an indirect correlation between linked *markers* and a phenotype, one can look directly at the effect of a particular *gene variant* on a phenotype. As a result, the power to detect subtle genetic effects is much higher, extending to effects that are undetectable by family studies. It is also much easier to find and recruit individual patients and controls than it is to collect extended families segregating a disease. The results of association studies—in particular, measures of risk conferred by a gene variant—are more directly applicable to the general population.

Association Studies: Interacting Genes

A polymorphism at one locus may have no effect on susceptibility by itself, but it may have a significant effect when coupled with a particular polymorphism at another locus. An illustration is provided by the system of enzymes that metabolize drugs, carcinogens, and other foreign chemicals. This metabolism is accomplished in two phases. First, a phase I enzyme attaches an oxygen to the chemical to form a reactive intermediate. Next, a phase II enzyme uses this oxygen to add a soluble group, which allows the removal of the chemical from the body. The intermediate compound is often toxic or carcinogenic. A mismatch in the rates of the two enzymes can lead to an accumulation of this compound, with obvious adverse consequences. For example, there is evidence that a particular allele of the phase I enzyme cytochrome P450 1A1, combined with a particular allele of the phase II enzyme glutathione-S-transferase, leads to a ninefold increase in lung cancer risk among Japanese smokers (Hayashi et al., 1993). Studies in mice have also shown that susceptibilities to both lung and colon cancer are influenced by interacting genes (Fijneman et al., 1996).

More complex interactions among multiple loci are also possible, and there is every reason to expect that they are the rule rather than the exception. How can we detect such effects? One approach is to examine directly the phenotypic effects of all possible allelic combinations at sets of loci. Although this may be feasible (if challenging) for pairs of loci, the number of combinations quickly becomes intractable when we consider larger sets.

An alternative is to look for correlations among polymorphisms in the general population. Selection over evolutionary time scales is likely to have favored particular combinations of polymorphisms at loci that jointly affect a phenotype, whereas variation at loci with independent phenotypic effects should be uncorrelated. The task is to uncover meaningful correlations among polymorphisms by analyzing the frequency of their joint occurrence in many individuals. Sets of interacting polymorphisms discovered in this fashion can then be tested for specific phenotypic effects. If successful, this two-stage approach offers greater power to detect subtle genetic effects by drastically limiting the number of combinations tested in the second stage.

Association Studies: Technical Considerations

Study Size

Standard statistical methods can be used to show that a sample of roughly 100 cases and 100 controls is sufficient to detect common polymorphisms that increase the risk of a disease fourfold—roughly the effect of the E4 allele on Alzheimer's—whereas a sample of about 1,000 is needed for polymorphisms that increase the risk twofold. Lesser risks require increasingly large samples, and practical constraints on sample collection and molecular tests will determine how small an effect can be detected. It is interesting to ask: What gene effects on phenotypes are too subtle to be detected even given the complete DNA sequence of every person on the planet?

Number of Alleles

Effect size depends on the number of functional alleles at a locus. Even if a locus contributes strongly to a phenotype, this contribution may be split among so many different polymorphisms that the effect of any one allele is undetectable. It is thus important to ask under what circumstances we can expect reduced allelic complexity. For instance, are there many fewer alleles in genetically isolated populations? This question is as yet largely unexplored.

Multiple Testing

When many polymorphisms at many loci are tested against many phenotypes, we run the risk of being swamped by false-positive results that will inevitably occur in such a fishing expedition. How do we combat this problem? There are indications that quite strict statistical criteria can be applied without too onerous an increase in sample size, but more precise guidelines need to be formulated.

Population Substructure

False associations will also show up when both a disease and a polymorphism have different frequencies in different ethnic subgroups of a population. Such substructure can be a problem even within broad racial categories such as "Caucasian." The problem can be overcome by using family members as controls, but remains an issue for data sets in which family members are not available. This is the case with many valuable population-based samples that have been collected over the years, for example the Framingham study.

Summary

Association studies directly address the genetic basis of disease in populations. They are quite powerful and require relatively few assumptions. Ultimately, this approach will dominate in studies of the relationship between genotypes and phenotypes. In the shorter term, the approach faces technical challenges. A systematic search of the genome by direct association involves tests of up to several variants in each of roughly 100,000 genes. It also requires the variants to be known, which may be difficult in the case of noncoding polymorphisms that alter the expression level of a gene by changing a control region or by interfering with the processing of messenger RNA. Until all variants are known and it is feasible to test them, studies based on ancestral signatures of common mutations may offer a shortcut.

GENETIC BASIS OF DISEASE IN POPULATIONS: ANCESTRAL SIGNATURES

Although some diseases such as hemophilia are frequently caused by new mutations, many important disease mutations arose only once or a few times in human history. For example, although many mutations in the cystic fibrosis transmembrane conductance regulator gene can cause cystic fibrosis, a single mutation of unique origin, known as ΔF508, is responsible for 75 percent of the cases. Similarly, the sickle cell mutation in the hemoglobin gene is likely to have arisen between one and four times. We know this because of ancestral signatures in the DNA surrounding a mutation.

Every mutation must arise at a particular time in a particular person—some baby born in Africa thousands of years ago carried the first sickle cell mutation. A mutation arises on a particular chromosome, with a unique signature of spelling changes. As generations pass, the signature is eroded by recombination, but remains largely intact in a region immediately surrounding the mutation, as illustrated in Figure 2.

Such a signature can serve as a proxy for knowing the mutation itself and can be tested for its phenotypic effect. This approach, known as linkage disequilibrium (LD) mapping, has been instrumental in narrowing the location of a number of Mendelian disease genes, including those for cystic fibrosis (Kerem et al., 1989) and diastrophic dysplasia (Hastbacka et al., 1992). The use of LD mapping for common diseases, as well as for initial gene detection, is largely untested. This situation is expected to change shortly as new technologies begin to allow systematic searches of the genome for ancestral signatures.

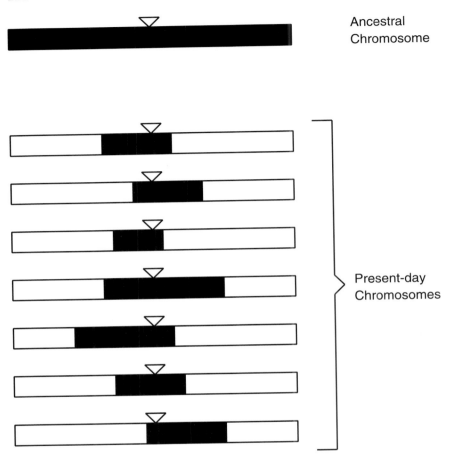

FIGURE 2 Signature of an ancestral mutation in present-day chromosomes. The ancestral mutation is denoted by the triangles. Shaded segments of present-day chromosomes retain the ancestral signature.

Example: Familial Mediterranean Fever

Familial Mediterranean fever (FMF) is, as the name suggests, a disease of inflammation that runs in families of Mediterranean origin. FMF patients suffer from periodic bouts of fever accompanied by other symptoms. Although FMF is rare in most populations, its frequency reaches 1 percent among non-Ashkenazi Jews, Armenians, Arabs, and Turks. This observation raises interesting questions about a possible selective advantage for carriers in the Mediterranean region and what the selective agent might be.

FMF shows simple recessive inheritance and is particularly common in offspring of consanguineous marriages. Working with Dan Kastner and colleagues, we used linkage to localize the FMF gene to a region of roughly four million base pairs on chromosome 16. Although this stretch of DNA comprises only a tenth of 1 percent of the entire genome, it is still too large for a systematic gene search. To narrow the location further, we searched for an ancestral signature in the North African Jewish population. The pattern of spelling changes in the DNA of patient chromosomes indicated that over 90 percent of them descend from a single ancestral chromosome on which the original FMF mutation arose. Lining up the ancestral segments (as illustrated in Figure 2), we were able to confine the gene's location to several hundred thousand base pairs (Levy et al., 1996), a region small enough to permit the identification of the gene, which was announced recently (International FMF Consortium, 1997). I return to the FMF story below.

Ancestral Signatures: Technical Considerations

Existence of Ancestral Signatures

LD between a functional polymorphism and surrounding markers must first be established in a population. It then needs to be maintained for many generations over appreciable chromosomal distances, without being eroded beyond recognition by recombination, mutation, and other factors. Under what circumstances are these conditions likely to be met, and how universal a phenomenon is LD? Except for young isolated populations, only limited theoretical guidance—and very little data—currently exists to begin answering this question.

Chromosomal Extent of Ancestral Signatures

Some mutations are old, arising when our ancestors wandered the plains of East Africa 100,000 years ago. Recombination will have had 5,000 generations to whittle away their ancestral signatures. Other mutations are more recent, arising in Finland 2,000 years ago, or being introduced into Canada by French settlers in the seventeenth century. These will be surrounded by much larger ancestral signatures. The density of genetic markers required to detect ancestral signatures thus depends on a mutation's history. Theoretical arguments can be used to estimate that a systematic search of the genome for LD in old mixed populations (such as that of the United States) will require roughly 500,000 markers (Kruglyak,

1999). Better empirical data are needed to accurately address this question.

Background Noise

Shared ancestral segments can occur by chance. The extent of such background sharing is not well characterized. A better description of this "noise" is essential for us to be able to recognize the signal of sharing related to phenotype.

Mutations with Multiple Origins

A particular mutation may arise more than once in the history of a population. It will then occur on chromosomes with several different ancestral signatures. For example, early-onset torsion dystonia is caused by a deletion of three nucleotides that appears to have occurred multiple times in different ethnic groups (Ozelius et al., 1997). In such cases, no correlation can be found between any one ancestral signature and disease. We need to have a better idea of how common this situation is.

Summary

An approach based on ancestral signatures has the advantage of detecting effects that are due to previously unknown variants. This approach would be very attractive if a reasonable number of markers sufficed to adequately cover the genome. However, as seen above, this is unlikely to be the case in general populations—around 500,000 markers may be required. The necessary number of markers can be smaller for young (recently founded) isolated populations, although only under a narrow set of conditions (Kruglyak, 1999). LD mapping also relies on a number of assumptions that are needed to ensure that ancestral signatures are established and maintained. Validating these assumptions requires greater knowledge of population structure and history—and their impact on allelic complexity of different populations and on the extent of ancestral signatures. Obtaining such knowledge is the subject of the next section.

POPULATION HISTORY
TRACES OF HUMAN EVOLUTION IN DNA

Example: Familial Mediterranean Fever

I now return to the story of FMF. As noted above, ancestral signatures played a key role in fine localization and eventual identification of the FMF gene. When the disease-causing mutations in the gene were identified, surprises emerged (International FMF Consortium, 1997). Over 80 percent of patient chromosomes carry one of four mutations, each changing a single amino acid. Patterns of spelling changes in the surrounding DNA strongly suggest that each mutation arose only once. Strikingly, patients of Armenian, Ashkenazi Jewish, and Druze origins share the same mutation and the surrounding ancestral signature, indicating common ancestry of populations that have been physically separated for over 2,000 years and were previously thought to be unrelated. Another mutation is found in Jewish patients of both North African and Iraqi origins, even though the latter group has been isolated from the rest of the Jewish population since the Babylonian captivity of 2,500 years ago. Similar unexpected relationships were found among other Mediterranean populations. We are currently extending these studies to learn more about the history and interrelatedness of the region's people.

Reconstructing History from Molecular Data

The example of FMF raises a general question: How do we reconstruct the history of human populations? Diversity of individual populations and relationships between populations are of inherent interest. As noted above, they also provide important information about the suitability of a population for finding disease genes.

We could begin the reconstruction by examining written records, oral histories, and mythologies. However, such records for most populations are incomplete at best, and even in the case of unusually well-characterized populations, such as Icelanders or the Amish, much of the information about population diversity still needs to be either discovered or confirmed empirically. If a population uses names that are passed from generation to generation in a systematic fashion, we could examine the number and frequency of different names. Such "last name analysis" has actually been used with success to assess the genetic diversity of some genetically isolated populations, including the Hutterites and French Canadians, but it cannot be used to establish relationships among populations or to shed light on the chromosomal extent of ancestral signatures.

The best repository of historical information is the DNA of population members. Patterns of spelling changes in the genetic code contain a record of population divergences, migrations, bottlenecks, replacements, and expansions. Indeed, molecular approaches have already contributed a great deal to our understanding of human evolution. A prominent example is provided by the studies of mitochondrial DNA that led to the "mitochondrial Eve" hypothesis of a recent African origin of modern humans. These approaches need to be updated to take full advantage of modern molecular tools, provide a more detailed look at the entire genome, and resolve finer points of population diversity and relationships. In the future, complete DNA sequences will be used for this purpose. In the meantime, we must rely on sampling polymorphisms from genomes of population members. Given such data, we can observe the number of different ancestral signatures present in our sample, their frequency distribution, and the range of chromosomal distances over which they extend. Such observations contain information about a number of historical parameters of the population, including its age, number of founders, rate of expansion, and bottlenecks (population crashes). For example, the presence of extended ancestral signatures indicates that a population is of recent origin, whereas the existence of only a few ancestral signatures indicates that a population has limited genetic diversity, likely due to a small number of founders or a bottleneck since the founding of the population.

Reconstructing History: Technical Considerations

Which Markers?

Several options are available. Microsatellite repeats, currently the most commonly used genetic markers, are highly polymorphic, but mutate at high frequency. Single nucleotide changes are highly abundant in the genome, more stable over time, but less polymorphic. Large insertions and deletions, most likely to represent unique mutational events in the history of the population, are relatively rare. These different types of markers may each have a role to play in answering different questions, depending on the relative importance of the polymorphism level (which allows us to distinguish different ancestral signatures), mutational stability (which allows us to look far back in time), and abundance (which may be needed to adequately sample the genome).

Marker Location

Most simply, markers can be chosen in random genomic regions, where they can be used to examine the extent of LD and the haplotype diversity in a population. Alternatively, they can be chosen around already known functional polymorphisms—such as those in the APOE gene associated with Alzheimer's disease—where their usefulness in detecting association can be tested.

Population Sample

Given an appropriate set of markers, we need to know how many members of the population must be included in the study sample, as well as how they should be chosen. One to two hundred independent chromosomes (or fifty to a hundred people) should allow a reasonably accurate estimation of population parameters, although the exact numbers will need to be computed for each particular task. The numbers will also vary from population to population, with more diverse, older populations likely to require larger samples for accurate characterization.

Summary

Reconstructing population history is one of the traditional goals of population genetics, and tools and insights from that field will undoubtedly be useful. The traditional approaches will need to be refocused from broad evolutionary questions about human origins to finer-scale population characteristics and relationships, particularly those that are relevant to searches for disease genes. They will also need to be updated to handle the amount and type of information that can be readily generated by modern molecular techniques. Finally, we can pose a provocative question: What aspects of human history cannot be reconstructed even from the complete DNA sequence of every person on the planet and are thus fundamentally unknowable (at least without a time machine)?

BEYOND DNA SEQUENCE: REGULATORY NETWORKS

DNA sequence is a static record of genetic information. The dynamic response of a cell to changing conditions is governed in large part by which genes are expressed—transcribed into messenger RNA. Understanding the regulation of gene expression is thus a key part of studying any important biological process.

Several emerging technologies now allow simultaneous measurements of transcription levels of many genes in a cell under a variety of conditions. It is then relatively straightforward to identify genes whose levels of expression under two conditions differ strongly. A more general—and difficult—task is to reconstruct the network of interactions among genes from these observations. I conjecture that this network will reveal distinct modules of genes whose expression levels correlate strongly within a module but more weakly between modules. Evidence of such modules has been found in a study of p53-mediated apoptosis (Polyak et al., 1997), as well as in a study of metabolic shift from fermentation to respiration in yeast (DeRisi et al., 1997). A particularly interesting question then arises: Do cells respond to changes in conditions simply by turning modules on and off or by reconfiguring the entire system in novel ways?

The budding yeast *Saccharomyces cerevisiae* provides a well-suited system for investigating this question, because the expression of essentially all genes can be monitored simultaneously under a broad range of precisely controlled growth conditions. Preliminary calculations suggest that a reasonable number of whole-genome expression measurements, each carried out under different growth conditions, suffices to untangle real correlations between pairs of genes from random noise. These pairwise correlations can then be examined further to uncover networks of interactions among larger collections of genes. I am currently planning computer simulations to test these ideas. I have also established collaborations with several yeast biologists to develop experiments aimed at addressing the question experimentally. Insights drawn from yeast can then be applied to more complex systems such as human tumorigenesis.

Dynamic responses of a cell are not confined to changes in the levels of gene expression. Many important processes are governed by modifications and interactions of proteins. Technologies to monitor levels of proteins, to distinguish different modifications of a protein, and to detect interactions between proteins are improving rapidly. Eventually, these technologies will allow a global examination of all proteins in the cell. Ideas developed to understand networks of interacting genes can then be applied to this next level of complexity.

THE ROLE OF THEORY

Theoretical thinking has a long intellectual tradition in biology. Reconciling the evolutionary ideas of Darwin with the Mendelian concept of particulate inheritance required the landmark quantitative efforts of Fischer, Wright, and Haldane. Theoreticians such as Delbruck, Crick, and Gamow made essential contributions in the early days of molecular biology (Judson, 1996). Today, as we face the challenge of understanding complex biological systems, theory has an important role to play.

Numerous complex systems are characterized by having many functional units with dense networks of interactions and by the need to infer underlying structure from incomplete, noisy observations. As a result, they pose conceptually similar problems. For example, the approach to understanding networks of gene expression described above draws on lessons and insights from work on information processing by large assemblies of neurons. The same ideas can be applied to understanding correlated genetic variation in populations.

A fruitful theoretical approach to complex systems needs to be firmly grounded in the data in order for the results to be applicable to the real world. At the same time, it should do more than simply react to the data. Indeed, the most valuable contribution of theory is to guide the field by making predictions and suggesting novel experiments.

The cultures of research, and not just techniques, differ between fields.

Delbruck, Crick, and Gamow—physicists entering molecular biology—brought with them a spirit of free information exchange, infectious enthusiasm, and easy speculation, evident in the informal scientific networks of the Phage Group and the RNA Tie Club (Judson, 1996). I hope to foster a similar intellectual atmosphere in modern human genetics, both within my research group and in the broader research community.

EDUCATION: A KEY RESPONSIBILITY

Public Perception of Genetic Research

The current level of public understanding of genetics is distressingly low. This is both contributed to and illustrated by coverage in the popular media—with headlines such as "Your whole life is in your genes at birth" from the tabloid *World Weekly News*, "Gene mapping: A journey for tremendous good—or unspeakable horrors" from *The Village Voice*, and "Couples can order embryos made to their specifications" from the *New York Times*. The recent science fiction film *GATTACA* is set in a "not-too-distant future" world in which every child's likely fate, including the time and cause of death, is predicted at birth from DNA analysis. Such portrayals tend to overstate greatly the degree to which genes determine physiology and behavior. They also frequently focus on extreme examples of genetic technologies, such as human cloning. As a result, they run the twin risks of, on the one hand, a backlash against all applications of genetics, no matter how beneficial, and on the other hand, a lack of preparation for the hard choices that lie ahead.

The Role of Education

I believe that our most important responsibility in dealing with the consequences of our research is education. Scientists cannot make choices for society—although we can, and should, avoid and discourage lines of research that we find unethical. Instead, we must make our research and its implications clearly and broadly understood. The public needs to know that we will never be able to make accurate predictions based on DNA sequence about many important aspects of human life, such as time and cause of death, level of accomplishment, and happiness. Similarly, although the medical benefits of human genetics are considerable, they should not be oversold, especially in the short term. Progress in understanding a disease—and especially in developing a treatment—can take many years, even after we know its genetic basis. Excessive emphasis on the medical benefits also short-changes the purely intellectual rewards of greater knowledge of human biology and evolutionary history.

Armed with appropriate information, society will be better prepared to face the complex questions posed by progress in human genetics, reaping the benefits while avoiding the pitfalls.

CONCLUSION

As human genetics enters the twenty-first century, many powerful molecular tools are either in place or about to arrive. They include a large capacity for DNA sequencing, an ability to examine many polymorphisms in many people, and techniques for monitoring expression levels of many genes under many conditions. These tools are already beginning to generate an avalanche of data. Early in this century, we can envision obtaining complete DNA sequences of many individuals, expression levels of all genes in a variety of cells, and a complete characterization of protein levels and interactions. The key task facing human genetics is interpreting this data.

In this chapter, I have tried to sketch out some of the intellectual tools that are needed for this task. The description is certainly not exhaustive, and some needs cannot be anticipated today. It is instructive to recall that the polymerase chain reaction is just over a decade old, but has had a profound impact on virtually all aspects of molecular genetics—it is difficult to imagine how most current approaches would be feasible without it. New technologies of similar power are certain to come along. Flexibility and openness to new ideas are thus of the essence. A unifying thread is provided by the realization that quantitative thinking will always be a necessary complement for experiments carried out on a global genomic scale.

Success in the task of interpretation will clarify the genetic basis of human diseases and other phenotypes. It will also provide insights into interesting aspects of population history. The general approach extends beyond human genetics to studies of pathogens and organisms of agricultural importance, and even more broadly to a better understanding of genetics and biology of all organisms. Thus, the ultimate benefits to society take many forms: better medicine, improved quality of life, increased knowledge, and a greater understanding of who we are and where we fit in among other forms of life.

REFERENCES

Botstein, D., D. L. White, M. Skolnick, and R. W. Davis. 1980. Construction of a genetic linkage map in man using restriction fragment length polymorphisms. *American Journal of Human Genetics* 32:314–331.

DeRisi, J. L., V. R. Iyer, and P. O. Brown. 1997. Exploring the metabolic and genetic control of gene expression on a genomic scale. *Science* 278:680–686.

Fijneman, R. J., S. S. de Vries, R. C. Jansen, and P. Demant. 1996. Complex interactions of new quantitative trait loci, Sluc1, Sluc2, Sluc3, and Sluc4, that influence the susceptibility to lung cancer in the mouse [see comments]. *Nature Genetics* 14(4):465–467.

Hastbacka, J., A. de la Chapelle, I. Kaitila, P. Sistonen, A. Weaver, and E. Lander. 1992. Linkage disequilibrium mapping in isolated founder populations: Diastrophic dysplasia in Finland. *Nature Genetics* 2:204–211.

Hayashi, S. I., J. Watanabe, and K. Kawajiri. 1993. High susceptibility to lung cancer analyzed in terms of combined genotypes of P450IA1 and mu-class glutathione S-transferase genes. *Japanese Journal of Cancer Research* 83:866–870.

International FMF Consortium. 1997. Ancient missense mutations in a new member of the RoRet gene family are likely to cause familial Mediterranean fever. *Cell* 90:797–807.

Judson, H. F. 1996. *The Eighth Day of Creation*, expanded ed. Cold Spring Harbor, N.Y.: CSHL Press.

Kerem, B., J. M. Rommens, J. A. Buchanan, D. Markiewicz, T. K. Cox, A. Chakravarti, M. Buchwald, and L. C. Tsui. 1989. Identification of the cystic fibrosis gene: Genetic analysis. *Science* 245(4922):1073–1080.

Kruglyak, L. 1999. Prospects for whole-genome linkage disequilibrium mapping of common disease genes. *Nature Genetics* 22:139–144.

Kruglyak, L. and E. S. Lander. 1995. Complete multipoint sib pair analysis of qualitative and quantitative traits. *American Journal of Human Genetics* 57:439–454.

Kruglyak, L. and E. S. Lander. 1998. Faster multipoint linkage analysis using Fourier transforms. *Journal of Computational Biology* 5:1–7.

Kruglyak, L., M. J. Daly, and E. S. Lander. 1995. Rapid multipoint linkage analysis of recessive traits in nuclear families, including homozygosity mapping. *American Journal of Human Genetics* 56:519–527.

Kruglyak, L., M. J. Daly, M. P. Reeve-Daly, and E. S. Lander. 1996. Parametric and nonparametric linkage analysis: A unified multipoint approach. *American Journal of Human Genetics* 58:1347–1363.

Lander, E., and L. Kruglyak. 1995. Genetic dissection of complex traits: guidelines for interpreting and reporting linkage results. *Nature Genetics* 11(3):241-247.

Levy, E. N., Y. Shen, A. Kupelian, L. Kruglyak, I. Aksentijevich, E. Pras, Je Balow, Jr., B. Linzer, X. Chen, D. A. Shelton, D. Gumucio, M. Pras, M. Shohat, J. I. Rotter, N. Fischel-Ghodsian, R. I. Richards, and D. L. Kastner. 1996. Linkage disequilibrium mapping places the gene causing familial Mediterranean fever close to D16S246. *American Journal of Human Genetics* 58(3):523–534.

Ozelius, L. J., J. W. Hewett, C. E. Page, S. B. Bressman, P. L. Kramer, C. Shalish, D. de Leon, M. F. Brin, D. Raymond, D. P. Corey, S. Fahn, N. J. Risch, A. J. Buckler, J. F. Gusella, and X. O. Breakefield. 1997. The early-onset torsion dystonia gene (DYT1) encodes an ATP-binding protein. *Nature Genetics* 17:40–48.

Polyak, K., Y. Xia, J. L. Zweier, K. W. Kinzler, and B. Vogelstein. 1997. A model for p53-induced apoptosis. *Nature* 389:300–305.

10

The Body in Parts:
Disease and the Biomedical Sciences
in the Twentieth Century

Keith A. Wailoo

SCIENCE, HISTORY, AND THE PROBLEM OF DISEASE

This chapter focuses not on genes per se, but on the history of genetics itself. Over the past quarter-century, many historians of science and medicine have engaged in the study of such scientific disciplines and of the diseases scrutinized by these discipline (Fox and Fee, 1990; Rosenberg and Golden, 1992; see also Bates, 1992; Bayer, 1987; Brandt, 1985; Brumberg, 1988; Jones, 1981; Leavitt, 1996; Rogers, 1992; Rosenberg, 1962; Rosner and Markowitz, 1991). For historians, the biomedical sciences are themselves important historical formations—always changing, especially as their tools and objects of study have changed. As biomedicine itself has become an important force in our society, historians of biomedicine have become increasingly interested in the intellectual and technical developments that shape what scientists themselves say about disease. This historical endeavor has taken us, of course, into the realm of particular disciplines, such as genetics, immunology, virology, and so on. But as the very character of biomedicine has changed, our historical studies have taken us also into the history of society, of social movements, and of politics.

Thus, the recent history of oncology and breast cancer could not be written without understanding the significance of the women's movement in the late 1960s and early 1970s. The history of immunology, virology,

Department of Social Medicine and Department of History, University of North Carolina at Chapel Hill and AIDS could not be written without attention to the history of gays in American culture. And the history of cancers like leukemia could not be written without attention to the prominence of children and catastrophic children's diseases after World War II; and the history of prostate cancer could not be written without attention to the demographic history of men and their aging in late twentieth century America. The history of disease in this century also takes us increasingly, therefore, into the political realm—into the realm of disease-specific legislation, policy making, and activism. And finally, the history of disease introduces us not only to science and the history of scientific theories of disease, but also to the history of technology. By the history of technology, I mean to focus on new drugs and new diagnostic technologies like the HIV test, on the ways they are used to shape public health agendas, and on the economics of how these technologies are used in the context of managed care. These are the kinds of concerns—the intellectual, the social, the political, and the cultural concerns–that shape the work I do as a historian of medicine, a historian of the biomedical sciences, and a historian of disease.

Using this kind of model—this very synthetic, complex model of disease—historians and other scholars have been writing fascinating new histories over the past 20–30 years. Our knowledge of disease is expanding dramatically through these histories. I can cite, for example, a number of examples of innovative histories. The history of anorexia nervosa, written by Brumberg (1988), reveals the entanglement of this eating disorder with the history of diet, nutrition, and also the history of body image among adolescent girls and young women in American culture. We see in recent histories of AIDS that the disease can be studied in particular cultural contexts. These studies—such as Farmer's (1992) study of AIDS and Haiti—make us aware that the face of disease varies not only from one time to the next, but also from place to place (Farmer, 1999). And historians of medicine and the biomedical sciences have not been afraid to tackle controversial issues. Bayer (1987), for example, has explored the ways in which the very definitions of disease have changed with moral concerns in America. His study examines the ways in which homosexuality was viewed as a "disease"—and particularly how it first appeared as a "disease" in the *Diagnostic and Statistical Manual* of the American Psychiatric Association in the early 1950s, only to be removed in the 1970s in a very different cultural climate. The story of this disease's evolution is linked, of course, to patterns of psychiatric categorization and to changing social mores.

It might be said that these types of disorders—anorexia nervosa, homosexuality, and AIDS—because they are so controversial, are particularly vulnerable to being shaped by cultural forces. My work, however, examines diseases and medical fields that we might regard as not so vulnerable to cultural influences—the cancers, immune system disorders and immunology, the definition and management of pain, etc.

Diabetes, however, provides an excellent case in point of the complexity of the historical evolution of disease (Bliss, 1982; see also Feudtner, 1995). After the discovery of insulin in the early 1920s, one of the ironies of this sort of medical progress has been the transformation of diabetes from an acute, fatal disease into a chronic, fatal disease—a disease that has now become a lived experience. After insulin in the 1920s came penicillin in the wake of World War II, and these two technological therapeutic developments transformed diabetes into what we now recognize as a chronic disease—a disease that affects a far larger group of people than at any other time in human history. Thus, one of the ironies of technological development—from the standpoint of the historian of science and medicine—is the role of therapeutic intervention in solving old problems and creating new problems. Technologies have literally altered the face of disease, bringing a kind of ironic progress in the fight against disease (Plough, 1986; see also Fox and Swazey, 1992). More people live now with diabetes and die with diabetes than ever before—because of insulin, penicillin, and the ability to treat infections and, more recently, the advent of kidney dialysis to treat late-stage renal failure in diabetes patients (Peitzmann, 1992).

The story of diabetes can be taken as an archetypal story of disease in the twentieth century. We have seen the dramatic decline of infectious diseases that predominantly affected children—including diphtheria, tuberculosis, pneumonia—and, in the wake of that decline, we have seen the rise of new diseases associated with (1) medical progress, (2) aging of the population, and (3) increasingly higher cultural standards regarding health and well-being. This sort of ironic progress was captured in a joke by a colleague of mine at the University of North Carolina at Chapel Hill. He told me the anecdote of two men arguing about which of their people came from more culturally superior stock. Amassing evidence for their superiority, they went back and forth. Finally, one man said to the other, "When your people were still living in caves and painting themselves blue, my people already had diabetes." This is the sort of progress I believe we can witness in the history of disease in the twentieth century.[1]

My concern as an historian of the biomedical sciences is not with the history of disease alone, but with a particular relationship that has shaped that history. Particularly, my research will focus on the triangular relationship among science, the human experience and understanding of disease, and clinical practices. By "the science of disease" I mean the emergence and transformation of models of disease as they have been developed in various scientific disciplines. One aspect of this research project involves investigating how science has, or has not, influenced how patients think about their diseases. This aspect of the research examines the relationship between scientific tools and theories and the individual, cultural, political, and social meanings of disease throughout the century. Another aspect of the research explores how these two ways of thinking about disease—the human experience and the scientific understanding— have, or have not, informed clinical practice from one time to the next.[2] This is primarily the model I have used both in my previous work and in the model that I plan to use in the pursuit of the James S. McDonnell Foundation Centennial Fellowship Program.

My previous work has focused on blood and ideas about blood disease. My first book, *Drawing Blood: Technology and Disease Identity in Twentieth Century America*, was not simply a study of blood diseases that have become more prevalent (Wailoo, 1997a). It also involved the study of diseases that have disappeared over time. (It is a curious fact about historians that we like to study things that no longer exist, for they reveal as much as does the study of ideas or things that live on.) One of the diseases in my study was a disorder called "splenic anemia," a disease that came into existence in the 1890s but which was gone by the 1930s (Wailoo, 1997b). The question I asked was, "how could a disease come into existence and then disappear 40 years later?" To study that problem, I looked at the people who were writing about splenic anemia, and they were, by and large, abdominal surgeons. What I noted was that this was an era when abdominal surgery was coming into its own. In the wake of antiseptic surgery in the late nineteenth century, the human abdomen became something that surgeons could actually venture into with the possibility of solving problems—rather than creating infections that would kill their patients. In the wake of antiseptic surgery (and the lowered risk of mortality from abdominal operations), surgeons began to theorize about the way that certain organs worked. One of these organs was the spleen, which previously was a mysterious organ that seemed to have no particular function. Surgeons were called upon at this time to remove many injured spleens, and they observed that removal of the spleen resulted often in a rise in the red blood cell count. They theorized that injured or enlarged spleens might be responsible for low red blood cell counts, thus inventing a new disease concept—splenic anemia. They argued that removal of the spleen could be curative in these cases of "splenic anemia."

This way of thinking about a new disease gained prevalence in the early twentieth century and then disappeared for a number of reasons in the 1920s and 1930s. One of the reasons it disappeared was because some of the colleagues of this generation of surgeons began to question this kind of dubious definition of disease, and (moreover) internists and pathologists questioned the surgeons' diagnostic practices as well as their optimistic assessment of their rates of "cure." This kind of scrutiny of

surgeons became especially rigorous in the changing hospital environment—a more corporate workplace where diagnosis was no longer a matter for individual authorities, but was now a corporate (group) activity
involving a number of specialists. With these changes in the hospital
work, the disease "splenic anemia" gradually disappeared. It did so, according to one renown surgeon, not because it did not exist, but because it
was not allowed to exist—that is, all the patients who surgeons would
diagnose with splenic anemia were now being diagnosed with other disorders by other specialists. And surgeons were simply not given the kind
of freedom to operate that they had been given in the first two decades of
the century.

Such stories of how a disease moves through time in a clinical context
are complicated. They are entangled with clinical politics, with the changing science of disease as practiced by different specialists, with speculation and theorizing, and (as discussed below) with the history of patients
and changing attitudes toward patients. It is this particular relationship—
among science, medicine, and patients' perspectives on disease—that I
seek to explore through this project.

The history of the patients' perspectives on disease takes us into the
realm of trying to understand individual experience of illness, culture attitudes, as well as patients' social and political concerns regarding health
care. To explore the scientific perspective, of course, we must learn more
about the history of particular disciplines, about disciplinary debates, and
about the theoretical models and technologies used by particular disciplines to shape their understanding of disease. Finally, the project
explores how all of this is applied (or not applied) in concrete clinical
settings, shaped by the style or politics of specialties such as surgery, pediatrics, obstetrics, and so on. The goal of the study is to examine how
this interaction has worked in the course of the twentieth century and
how this interaction has shaped the history of disease in this century.

A current example of this dynamic reveals the continuing complexity
of this relationship of science, medicine, and human experience of illness.
Just a few years ago, the discovery of the BRCA1 gene for breast cancer
began to transform the ways in which patients and clinicians as well as
scientists thought about what breast cancer is and how we should respond
to it. One of the results of the discovery of the BRCA1 gene has been an
increased awareness of the hereditary nature of some breast cancers (indeed a tiny number of breast cancers). Inheritance of the BRCA1 gene
explains why only a tiny percentage of women develop breast cancer. Yet
the level of cultural sensitivity to new findings in genetics is so high that
such findings have been widely disbursed. The spread of information
about the "inheritance of breast cancer" has produced confusion rather
than clarification of thinking about breast cancer. As one recent article in

the *New York Times* notes, it has produced a combination of new data, hope, fear, and confusion (Grade, 1999). The *Times* article was based on a clinical study that suggested that many, many women are convinced that breast cancer is hereditary—many more women than ought to believe it. And this perception has important clinical and human consequences. Many women and their physicians misread the message of science, believe that the presence of the BRCA1 gene determines that cancer will appear, and increasingly opt for "preventive mastectomies" in order to safeguard against a cancer that has not yet (and may not) appear. The popularity of the belief in the inheritance of breast cancer and genetic testing has much to do with the importance of genetics as a biomedical specialty in our time (not unlike the importance of abdominal surgery in the early twentieth century). Indeed, if there is a general conclusion to be drawn here it is that even at the end of the century, the interaction among scientific findings, human perceptions, and clinical practices continues to be complicated and problematic.

My project involves an eight-year study of four different categories of disease, disorders, and illness experiences—cancers, genetic diseases, disorders associated with the immune system (from smallpox to polio to AIDS), and the problem of pain (Mann, 1988; Pernick, 1985; Rey, 1995). Within these vast areas, I have chosen to write the history of particular disorders—a task that I believe can capture some of the complexity of the history of the fields themselves. The goal at the end of the project is to produce a synthetic study of disease in the biomedical sciences in the twentieth century that can speak not only to patients' experiences as they have evolved, but a history that can inform contemporary clinical debates and can inform science policy as well. In the remainder of this chapter, I discuss the history of one genetic disease—cystic fibrosis—and present a model for how this project will unravel the history and evolution of particular diseases.

DISEASE AND THE BIOMEDICAL SCIENCES
IN CULTURAL CONTEXT

Because cystic fibrosis (CF) is one among several genetic disorders and fields that will be a focus on this project, it may be useful to say a few words about some of the other areas under study. Before describing its historical evolution as a disease, it might be useful to say a word or two about why these four areas—genetic disease, immunology, cancer, and pain—have been selected for this project. These particular areas, I argue, illuminated the transformation of disease in the twentieth century. In their different ways, they highlight some of the major shifts of the past 100 years—such as the decline of acute infectious diseases associated with childhood and the rise (almost in the wake of that decline) of new diseases that were previously invisible and unseen. Among these new diseases were genetic disorders such as CF and sickle cell disease. Their rising profile in clinics across America began to occur in the post-World War II era. Also revealed in the history of these disorders is the gradual aging of the population, a process that has brought about higher prevalence of many cancers (for example).

Each of the cancers at the center of this study, however, tells its own story. Childhood leukemia and breast cancer, of course, each tell stories about the illnesses and suffering of different segments of the population and how their particular suffering has become visible in scientific, clinical, and popular culture. The visibility of these diseases is shaped not only by the visibility of those groups affected—women and children—but also by new technologies. For example, radiation and chemotherapy after World War II were direct results of the wartime research on mustard gas and radiation. In the war's wake, these tools transformed medical institutions, creating new institutions like St. Jude's Children's Research Hospital in Memphis, Tennessee. The new research hospital symbolized the high hopes that Americans invested in these new tools. The hospital also symbolized the new visibility of previously obscure childhood diseases, in this case leukemia. This was all happening in the context of the post-World War II "baby boom," at a time in which Jonas Salk was pursuing the vaccine for polio, in which parents were signing up their children in droves for field trials despite the fact that no real serious trials had ever been done on the polio vaccine (Smith, 1990; see also Maier, 1972; Meldrum, 1994). (Indeed, the day that Salk's field trials began was called "V-Day," and the children who participated were handed certificates of bravery to reward their participation. There was much talk of heroism, wrapping the children in the atmosphere of wartime triumph.) It was a time of enormous faith among middle-class Americans in the science and technology that were used to solve human problems (a kind of faith that we do not see today). It is this complex of forces—the new technologies of chemotherapy and radiation, public faith in science, and the symbolism of childhood suffering—that explains why childhood leukemia rose dramatically in social prominence and why many Americans focused on this particular disease and on the new science of oncology during this era.[3]

In the case of immunology (and particularly the history of vaccination and transplantation), we find a different kind of clinical science—with a different history and social significance.[4] We find a clinical science that has been shaped overwhelmingly by technical innovations such as vaccination and transplantation and by controversies surrounding these innovations. Immunology was defined by vaccines and antitoxins in the late nineteenth and early twentieth centuries (some were successful and others failed) for disorders like diphtheria, tuberculosis, as well as polio-myelitis. The popular support for Salk's vaccine stands in sharp contrast to a much more checkered history of skepticism and public ambivalence about vaccination. One important reason for this difference is that vaccination depends on a particular kind of faith—a willingness to believe that intervention will protect a healthy individual from a future of disease. Thus, the study of vaccination—from the early twentieth century through the age of polio, and into the era of controversy surrounding a potential AIDS vaccine—allows us to explore not only the transformation of a science and the transformation of disease problems, this history also allows us to examine the ways in which immunology has intersected with matters of individual belief and public faith.

Each of these fields, then, reflects an important transformation in science, technology, culture, and medicine in the twentieth century. Pain—the fourth area under consideration—seems to sit apart from these other areas, but its study also reveals particular features of the biomedical sciences in the twentieth century. Pain is not disease itself, but a diffuse set of experiences—some of which can be associated with disease. Pain is a clinical problem and an ongoing scientific dilemma that changes with the disease in question. But the scientific understanding of pain and the clinical approach to pain has also changed over time, varied according to context, and has been shaped by cultural attitudes.[5] Pain relief in the context of American society is also big business—and understanding much more about the economics of disease management (a major theme in twentieth century biomedicine) will be facilitated by the focus on pain. Indeed, all of the problems I have chosen to highlight in this study—cancer, genetic disease, pain, and immunology—are areas where one sees pharmaceutical production and popular consumption intertwined. The practice of speeding relief along to the patient has been a dominant theme in American medicine, and this practice has also been important in shaping scientific ideas about disease, in shaping clinical practices, and in shaping how patients think about their disorders.

In this eight-year project, the first two years will be dedicated to studying the cancers, the second two-year period will be devoted to studying pain, the third two-year period will focus on immunology, and the final segment moves to the history of genetic disease. The research process will bring together scientists, clinical practitioners, clinical researchers, policy makers, and historians to talk about not only this history, but also the ways in which these histories are relevant to their practices today. This research project is an effort to generate a richer discussion about what we mean when we say we are pursuing "cures" for disease. What has that meant in the past? How have science, medicine, and human experience combined to shape the pursuit of cures in the past, and how should they interact in the future?

CASE STUDY: THE TRANSFORMATION OF CYSTIC FIBROSIS

After these quite general statements, it might be useful to underscore these points through a specific case study in the history of genetic disease.[6] I focus here on the history of CF, a disorder that was identified as a discrete entity certainly before World War II, but which earned a particular level of visibility and prominence in the decades after the war. The history of this particular malady underscores many of the interactions I described above—the changing role of science in the clinic, the role of aging in reshaping patients' expectations, the business of therapeutics and how it shapes medical practice (sometimes in negative ways), and the ways in which cultural understanding of disease can shape the course of science and medicine. The presentation here stems from research conducted with a graduate student, Stephen Pemberton, at the University of North Carolina at Chapel Hill as part of a project funded by the National Institutes of Health (NIH) over the past three years (Wailoo and Pemberton, forthcoming).

Throughout most of its history, CF has been framed as a rare disease, as a complex therapeutic challenge for patients, for physicians, and for clinical scientists. In the late 1930s, Dorothy Anderson, one of the pioneers in CF research, saw the disease as "cystic fibrosis of the pancreas"— a disease, as she saw it, characterized by severe nutritional malabsorption in small children and as potentially treatable by dietary supplements and nutritional management (Anderson, 1938; see also Farber, 1944, 1945). The diagnosis relied on the identification of pancreatic insufficiency until the 1950s, when researcher di Sant' Agnese and colleagues (1953) first noted that the electrolytes in the sweat of CF patients were elevated by comparison with people without CF. During the first two decades of CF's cultural history, then, researchers typically believed that malabsorption problems were the primary clinical features of the disease, and that this problem was accompanied by secondary but severe problems of mucus overproduction in the pancreas, lungs, and other organs.

One of the important transformations in how clinicians saw CF was the advent of antibiotics, first for wartime use and then subsequently for use in civilian populations. The production of synthetic antibacterial agents in the 1940s and 1950s allowed physicians to tackle a wide range of infections that manifested themselves as pulmonary congestion, pneumonia, tuberculosis, and so on in CF patients. The decline of these disorders which affected CF children meant a kind of unmasking of CF as a "new" disease—more clinically visible in this era than ever before. As antibacterial agents were more widely produced by a burgeoning pharmaceutical industry and deployed in a wide range of clinical contexts, CF became characterized as a "great impersonator." The challenge to clinical medicine was to recognize and to unmask this multidimensional disorder that affected so many parts of the body, and which so easily mimicked asthma, bronchitis, as well as infectious disease. This theme—the proper diagnosis and recognition of CF—dominated the clinical and scientific literature of the 1950s and 1960s.

At the same time, however, writing on antibacterial therapy as early as 1951, three CF researchers (Garrard et al., 1951, p. 485) noted that the enlarged antibiotic armamentarium of the physician today was "a mixed blessing" and "a double-edged sword." (This is a limitation of antibiotics that we are still familiar with today.) As these researchers noted, "any long continued therapy with a single antibacterial agent invites the development of highly resistant organisms which may flourish in an environment rendered more favorable by the absence of susceptible bacteria." So from the outset, clinical scientists and physicians were well aware of the problem of antibacterial therapy—that it was a balancing act for every clinician between combating the infectious organism that colonized the thick mucus of the lungs and preventing the proliferation of more-resistant bacterial strains that resulted from antibiotic overuse. Indeed by 1967, some researchers could point out that overuse of antibiotic therapy had actually produced a new biological problem in CF. "There is little doubt," wrote one researcher, "that the establishment of new species of *Pseudomonas aeruginosa* in the respiratory tract is encouraged by the suppression of other bacteria by antibiotics" (Boxerbaum et al., 1972; Burns and May, 1968; Kulczycki et al., 1978). Nevertheless, as Isles et al. (1984) pointed out, through the 1960s and 1970s the trend was toward the use of more and more antibiotics (in increasing varieties because of the wealth of pharmaceutical agents on the market). This style of practice continued into the late 1970s.

Physicians involved in the care of CF patients developed a kind of individualism in clinical practice. Individual practitioners developed their own therapeutic styles. These styles often involved alternating among different kinds of antibiotic agents, carefully tilting and tailoring their interventions to keep these bacteria off balance, carefully weighing each new agent on the market, and trying to manage the increasing variety of antibacterial therapies that became available—wide spectrum, narrow spectrum, oral, intravenous, and so on. By the 1970s and early 1980s, however, some researchers began to rigorously scrutinize these practices.

In the view of some, the treatment of CF patients had become little more than a set of individual rituals—as opposed to a standardized science.

The emergence of this kind of skepticism about the previous generation's therapeutic practice, and the embrace of a more "scientific" approach, should be put in historical perspective. Why did this shift occur? It happened in part because of the growing frustration of clinicians with the ways in which antibacterial agents were being used, but there is more to the story than that. The disease itself had been transformed by the advent of antibacterial therapy, and the "new" CF posed new problems to clinicians—problems that they hoped would be solved by a turn to science. Not only had *Pseudomonas* infection emerged as a by-product of individualized treatment styles, but a demographic transformation among CF patients had occurred. The number of adolescents with CF had grown in number through the 1950s, 1960s, and 1970s. Children with CF were growing up; they were becoming adults with an intimate familiarity with the medical system. As a result, their concerns could be voiced more effectively, and they could become more contentious in the clinical arena. As one CF physician recently noted, adolescents are a little harder to manage clinically—as well as socially. The changing demographics of the disease offered a new social profile as well.

As I noted above, when CF first emerged it was understood as a multidimensional disorder affecting many different organs. Certainly the pulmonary problems were significant, but it was understood in the early years as a nutritional disorder. Very early in the 1950s, one well-known NIH CF researcher Paul di Sant' Agnese insisted that another way (indeed the best way) of understanding the disorder was to think of it as a metabolic disease. That is, it should be understood as a disorder stemming from a faulty metabolic system—in which an underlying inability to regulate the production of mucus throughout the body caused problems in the lungs, pancreas, gastrointestinal tract, and elsewhere. This shift in scientific thinking and terminology occurred at the same time that antibacterial management grew. Indeed, as shown below, there emerged an important tension between the scientist's understanding of the disease and the clinician's understanding of the primacy of lung deterioration in the lives of people with CF. This tension would be resolved not in scientific conferences or in clinical discussion, but, perhaps surprisingly, in the political arena—reflecting the rising social stakes surrounding such diseases.

In 1972, the U.S. Congress considered legislation to increase research funds in several areas: heart disease, blood disorders, and lung disease. Looking to the established scientific conception of CF as a complex, metabolic disease, representatives of the NIH and the Department of Health, Education and Welfare insisted that CF was a "metabolic disease," and as such did not fit any of the funding categories (DuValm, 1972, p. 86). The proper characterization of CF was particularly important because funding for regional pulmonary pediatric centers (where CF patients had found care in recent years) had been greatly reduced. But legislators echoed the pronouncements of NIH scientists, asserting that CF was in fact a general metabolic disorder and that the biochemical disturbance responsible for the disease was not confined to the lungs. According to normative scientific thinking, the way to study CF was to tackle the underlying metabolic disturbances.

This view conflicted, however, with the view of the Cystic Fibrosis Foundation (CFF) that had come into existence in the 1950s and had grown in significance in political power and in lobbying force by the 1970s (Barbero, 1972, p. 216). Sitting before Congress in 1972, its director argued that CF was unquestionably a lung disease. Well aware of the need for more research dollars to study CF and of the declining funding for pulmonary pediatric centers, the CFF director argued that CF was a lung disease because pulmonary failure was the principal cause of death for children with CF. His point was an important one, for it reflected the very different sensibilities of clinicians as opposed to clinical scientists. The most important feature of the disease, in his thinking, was the way in which it killed patients (the cause of mortality), and not the underlying biological mechanisms of disease. "Children's lung diseases, in which many contributing causes exist—genetic and non-genetic—are known, and are a major source of concern in this day." He appealed to Congress to include CF in the 1972 legislation especially because of the decline of the regional medical programs. He concluded his remarks by noting, "it would be unsound to separate cystic fibrosis out from any of these acts of legislation. It is a lung disease" (Barbero, 1972, p. 217).

His words and his disagreement with NIH scientists over the categorization of CF offer a fine example of a broader process at work. This tension provides insight into the ways in which politics and the search for funding reframed these scientific debates and how the politicization of such diseases (increasingly an issue within biomedical research in the 1960s and 1970s) shaped the ways in which experts thought about people with disease. For both NIH scientists and CFF representatives, CF was one single disease. But which aspect of the disease warranted the most attention? The answer depended on the broader social, economic, and political context. For the CFF, it was imperative that Congress recognize the pulmonary mortality of the disorder and acknowledge that, in the patient's and clinician's perspective, lung deterioration had become the principal cause of mortality and the primary concern—replacing mortality from infectious disease. The director's argument, however, failed to sway legislators, who turned for expert advice to NIH scientists who would be carrying out the research in question. CF was not included in the final legislation.

But in many ways, the CFF's representation of disease resonated much more than the scientific perspective with the public, with patients (who were now adolescents and young adults), and their families. Not only had life expectancy increased, but also the number of people receiving care had more than doubled in the period from 1965 to 1975. Indeed, the higher profile of CF resulted in President Richard Nixon holding a Rose Garden ceremony with a CF poster child and the CFF in order to raise awareness of the disorder (*New York Times*, 1972).

At this juncture, then, the clinical realities of CF were complex, and clinicians were somewhat uncertain about what the science of CF meant for their clinical concerns. CF patients presented with biological problems such as *Psuedomonas* infection that resulted from the individualized antibacterial therapies. That is, CF physicians had literally created new biological problems for their patients because of their complex antibacterial management, and they were not compelled to manage the consequences. A wealth of such drugs made this style of practice the norm. CF patients were living longer, treated by a wider range of antibiotics, physiotherapy techniques, and other therapeutic modalities, but still dying from their lung deterioration. The standard laboratory-based, scientific understanding of the disease, however, did not stress this aspect of the problem. Rather, scientists stressed the need for studies of the basic underlying metabolic disturbances that cause mucus overproduction as well as the deterioration of lungs and other organs. Faced with a growing population of CF patients, clinicians also became aware of large variations among these patients—in the ways their disease actually appeared from one case to the next. One crucial problem seized upon by clinicians in the 1970s—and a route toward improving treatment—was to study whether any standard routines, based in science, could be applied to the care of CF patients? What was the actual efficacy of antibacterial therapies, especially in the context of these increasing variations in the disease biology and in clinical care. As one author noted in 1978, "while it is accepted by many that the increased longevity of the CF patient is strongly related to antibiotic use, this has never been adequately documented" (Kulczycki et al., 1978).

When we look at the clinical literature of the 1970s, we are stuck by a new development—the turn toward clinical research studies, placebo-controlled clinical trials, and other such studies in order to determine which of these antibacterial agents more effectively and efficiently controlled infections in CF (Beaudry et al., 1980; Hyatt et al., 1981; Kearns et al., 1982; Loening-Baucke et al., 1979; Nolan et al., 1982; Parry et al., 1977; Warwick, 1977). In these highly structured clinical trials, researchers explicitly questioned the impressionistic, ritualistic basis on which doctors in the past reported failures or successes in antibacterial treatment. What was the impulse behind this new science of CF? This research was guided, in some part, by the urge and need to manage the very excesses of pharmaceutical production. That is, there was a kind of embarrassment of riches in the realm of antibacterial therapy; the clinical scientists of the 1970s hoped to figure out how to rationalize the care of CF patients.

A "clinical science" emerged to address the complexities of medical care in the early 1970s. This science of CF stood in contrast to NIH science, for NIH researchers and others had seen CF as a test case for the study of metabolic disease. For them, an understanding of the relationship between pancreatic deficiency and mucoviscidosis would shed light on the physiology of metabolic diseases at large. But in this emerging clinical science, research on antibacterial agents could focus on the acute stage of CF lung failure, on the standardization of individualistic medical care, and on the problem of pharmaceutical abundance—significant problems for patients and practitioners. This was a very different kind of science than the science advocated by di Sant' Agnese because "clinical science" attempted to answer very specific, therapy-related, questions. How do we use drugs responsibly? Are there standard rules for managing the wealth of therapeutic options? In many ways, the story of CF at this juncture touched on fundamental questions regarding the responsible practice of science and the responsible use of science. Both of the sciences and the views of disease debated in this era were products of the time and context.

THE SPECULATIVE SCIENCE OF "GENE THERAPY"

The science of clinical trials to study antibacterial agents emerged from the clinical frustrations and social changes of the 1960s, but it also immediately predated another turn in the scientific understanding of CF—the rise of genetic identification and prospects for CF gene therapy. It is particularly important to place in proper historical and social perspective the new scientific models of CF that emerged from clinical medicine and genetics, beginning in the late 1970s and early 1980s. As a science that offers yet another definition of CF as a disease, gene therapy must be placed in its proper historical continuum if we are to fully understand and evaluate its claims. In many respects, "gene therapy" emerged itself from the frustration with clinical trials in the 1970s. Indeed, the clinical studies envisioned in the 1970s had failed to create standard models of CF care for one important reason—the disease itself resisted standardization in clinical care.

One reason that CF resisted standardization was the extreme variation and the complexity of the disease from one sufferer to the next. As one pediatrician noted in 1985, the issue of antibacterial management was still clouded and possibly unresolvable. He pointed to a number of reasons. Perhaps the variation in how CF actually manifested itself from one patient to the next made it impossible to hold to standard rules of antibacterial management. Faced with precisely these frustrations and at this point in the disease's history, genetics (and the gene therapy approach to CF) offered a new set of therapeutic possibilities. Pointing back to the deficiencies of basic science knowledge, one pediatrician noted, for example, "perhaps when the basic defect in cystic fibrosis is understood, the relationship between the host to the microorganism will be better understood" (Nelson, 1985). In the past 10–15 years, the possibilities of gene therapy have been enthusiastically embraced by many parts of the CF community—by pulmonologists, by clinical researchers, by patients and families, by the CFF, by biotechnology companies, and by many others. Indeed, one of the implicit promises of genetics and gene therapy was to enter this very frustrating atmosphere of clinical science and to provide a way out of lingering dilemmas in patient care and scientific understanding. Gene therapy, as the very name suggested, promised to make clinical science directly relevant to patient care. Since the early 1980s, gene therapy has attracted significant public and professional attention as a potential therapeutic modality in CF care and in other diseases. Much of the attention on gene therapy has been overly optimistic from this historian's point of view—especially for an experimental procedure in its early stages of testing that has produced no proven benefits for any single human patient with any disorder.

Speculation about altering genetic material to cure disease grew throughout the 1980s. But the advent of high optimism for gene therapy followed directly on the heels of the discovery in 1989 of one of the defective genes implicated as the cause of some cases of CF (Riordan et al., 1989; Seligmann and Glick, 1989). It was this missing gene that resulted in the failure of chloride to pass through the membranes of the cells of CF patients. Gene therapy was envisioned as a method of repairing the chloride transport process by inserting the repaired gene into the lungs. To understand and evaluate the science of "gene therapy," however, it is crucial to situate that science in its proper context—just as we situated the antibiotic clinical trials in their context. The science of gene therapy is a speculative, entrepreneurial science—a science shaped by a culture of economic risk taking, venture capital, and financial speculation. This is a very different science than the science of Dorothy Anderson, the science of NIH researcher Paul di Sant' Agnese, or the science of clinical trials in the 1970s and 1980s.

As early as November 1985, well before the discovery of the gene for

CF, *Business Week* magazine ran a cover story on the emergence of "gene doctors" who were portrayed as "erasing nature's mistakes" and "curing life's cruelest diseases" (Shulman, 1985; see also Ohlendorf, 1985). Other media pointed out that these new doctors were "closing in on CF" and "laying siege to the deadly gene" itself. Into the early 1990s, public, professional, and business enthusiasm for "gene therapy" as a potential cure for CF reached a peak as the successful testing of a new therapy seemed imminent (Brady, 1990; Nichols, 1993; Seligmann, 1990).[7] In early 1993, pulmonologist Ronald Crystal began a first experiment at the National Heart, Lung, and Blood Institute, "fulfilling hopes that had gathered steam like a locomotive force in the past several months," according to the *New York Times* (Angier, 1993b; see also Angier, 1992; Cowley, 1993). The study, one of three approved studies, involved a 23-year-old man with CF who, in the words of the newspaper, "inhaled the cold virus, the adenovirus, that had been altered to enclose a healthy copy of the CF gene the patient lacked." It would have been more precise for Crystal and the newspaper to label this a "gene transfer experiment" rather than "gene therapy" because the study was simply a test of the uptake of the viral vector and a study of its ability to alter chloride transport. The test was never envisioned as a test of the efficacy or a treatment. This was made clear elsewhere in the article, although the language of the article suggested that this was a therapeutic milestone.

At the same time, other pharmaceutical enterprises had turned to CF, sensing a lucrative marketplace of patients. Genetech, Inc. had requested Food and Drug Administration permission to market its CF drug, Dnase. Shortly after its approval, Genetech stock rose sharply by fifty cents a share. A third article from the same period pointed out, however, that Dnase was only one part of a "six hundred million dollar horse race," a race in which gene therapy was the much-celebrated, highly praised, but unproven front-runner (Hamilton, 1993; *New York Times*, 1993; see also Carey, 1993; *New York Times*, 1991). The key question was this: Who would capture this potentially huge market? Crystal himself was in the race. Indeed, as many articles noted, Crystal was cofounder of a gene therapy startup company named Genvec that had received $17 million in capital support from Genetech. Here was the scientist as entrepreneur in the business world of the 1990s. Indeed, many of Crystal's (and other "gene therapists'") vaunted predictions about the rapid development of gene therapy must be evaluated against this backdrop.

In late 1993, Crystal's experiments hit a technical and public relations snag. One research subject enrolled in his studies showed signs of lung inflammation, as well as drops in oxygen levels in the blood and evidence of pulmonary damage (Anglier, 1993a). These events provoked Crystal to speculate that this particular patient might have been idiosyncratic, or that perhaps the upper limits of adenovirus therapy had been found. For reporters, however, the problem underscored the severe difficulty of translating a highly speculative, experimental procedure into work-a-day clinical practice. Over the next few years, the early wild enthusiasm for "gene therapy" would begin to ebb. Predictions would be scaled back gradually, as geneticists, pulmonologists, and patients began to look somewhat more skeptically upon the enterprise. The unmitigated enthusiasm of the mid-1980s had given way to a willingness to characterize "gene therapy" for what it is—an experimental long-shot. In one survey in *Science* magazine in 1995 on "gene therapy's growing pains," the author noted that right from the start, gene therapists recognized that their central challenge would be technical ones: finding safe vectors capable of transporting genes efficiently into target cells and getting the cells to express the genes once they were inserted (Kolata, 1995; Marshall, 1995; Palca, 1994). These modest goals would define gene therapy in the late 1990s.

In this new environment in which commercial pressures no longer drove the high enthusiasm of scientists as much as it had earlier, researcher and director James Wilson of the Institute for Gene Therapy at the University of Pennsylvania could look back and comment that "this commercial pressure may also account for some of the hype surrounding developments in gene therapy" (quoted in Marshall, 1995). After all, he pointed out, if you are the leader of a gene therapy company "you try to put as positive spin on it as you can, on every step of the research process because you have to create promise out of what you have. That is your value." This was a very different type of science shaping the history of CF at the close of the twentieth century. Clinical science in this instance had become part of entrepreneurial development, and the very pronouncements about the promise and possibility of "gene therapy" for CF could be understood as part of the positioning of entrepreneurs for capital and market share (Baggot, 1998; Brown, 1995; Martin and Thomas, 1998).[8] The scientist–physician–businessman spoke highly of the promise of gene therapy, in large part, in order to generate enthusiasm and interest (Friedmann, 1994).

What conclusions can we make about the history of this disease, particularly when we ponder the relationship between the science of disease, the clinical management of disease, and the patients' perspectives on disease? Through tracing the history of this one disease, CF, I have made some inroads into this question. Indeed, the disease's history reveals a great deal about (1) the transformation of science; (2) the historical evolution and impact of clinical practice; and (3) important changes in the politics, the social reality, and the lived experience of disease in American society. Demographic trends in the aging of the CF population—a trend driven in large part by therapeutics—has brought a patient's perspective into public view. Science itself has produced many changing models of CF—as a nutritional disease, as a metabolic disease, as a lung disease, and as a genetic disease—and (as demonstrated above) each of these models has been shaped by technical developments and by the politics of the time. Each of these scientific models of disease has spoken to patients' and cultural perspectives in quite different ways. Finally, I have attempted to sketch how these sciences as well as the cultural history of CF have been shaped by (and shape) the clinical management of the disease and how clinical practices themselves—such as the use of antibacterial agents— have altered the biology of the disease and the patient's experience.

CONCLUSION

Let me conclude with a few observations from this history, conclusions that establish some of the parameters of my project as I turn to examining the histories of cancer, pain, immunology, and genetic disease. Diseases are evolving entities, changing with the interaction of science, technology, clinical practices, politics, and culture. When we speak of the responsible use of science, it is important that we understand the character of the science in question and understand its historical trajectory and its broader social entanglements. This kind of knowledge, I believe, contributes to an enriched and informed discussion about the future direction of science and society.

The history of CF is only one aspect of this James S. McDonnell Centennial Fellowship research project. This fellowship will allow me to purchase books, it will allow me to hire research assistants, and it will allow me to organize workshops and symposia to discuss and disseminate the findings of my research. The goal is to programmatically influence the kinds of discussions that happen in the clinical arena, in the scientific arena, and in the history of medicine and science regarding the problem of disease in our time. Envisioned in this project are a series of annual scholarly workshops, the first of which will be held in 2000 on cancer, possibly breast cancer. Such workshops will bring together breast cancer researchers, clinical scientists, policy makers, and historians to discuss the interaction of science, medicine, and the patient's disease experience. It is possible that the workshops will produce edited volumes on the history of cancer. Every other year the project will result in a public symposium to disseminate this research to a wider audience. It is possible that the workshops and symposia will lead to the production of papers and books that inform the public and mass media about how to evaluate the science of disease, or about understanding this dynamic, changing relationship among science, medicine, and society.

ACKNOWLEDGMENT

I thank the James S. McDonnell Foundation and the selection committee for the Centennial Fellowship in history and philosophy of science for their extraordinarily generous support for this research project.

NOTES

1. On the rise of chronic disease and the decline of acute infectious disease, see Boas (1940).

2. A wide range of scholarship has explored the complex relations among clinical practice, science, and disease experience. See Harvey (1981, p. xvii), Lederer (1995), Marks (1997), and Matthews (1995). For more on the development of the clinical sciences in their social and political contexts, see Beecher (1966), Booth (1993), Etheridge (1992), Harden (1986), and Rothman (1991). See also Beecher (1959), Bynum (1988), Katz (1972), and Pappworth (1967).

3. On the history of cancer, see Cantor (1993), Patterson (1987), Peller (1979), Rather (1978), and Strickland (1972). See also Bud (1978), Markle and Petersen (1980a, 1980b), Panem (1984), Peters (1993), Proctor (1995), Rettig (1977), Richards (1991), Sigerist (1932), Studer and Chubin (1980), and Young (1980). On the U.S. Public Health Service trials, see Endicott (1957), Fujimura (1996), Jacyna (1988), and Lowy (1996).

4. The literature on the history of immunology is vast. See, for example, Brent (1997), Hall (1997), Lowy (1989), Martin (1994), Moulin (1989a, 1989b), and Tauber (1994).

5. The literature on pain is also vast. See Mann (1988) and Pernick (1985). And for a statement from a twentieth century scientist, see Rey (1995), Unruh (1996), and Wall (1975). See also Friedlander (1992), Leavitt (1986), Mann and Plummer (1991), McTavish (1987), Meltzer (1990), Morris (1991), Nelson (1993), Papper (1992), Pitcock and Clark (1992), Rushman et al. (1996), Scarry (1985a, 1985b), Szasz (1957), Tweedie and Snowdon (1990), and Wear (1995). On the current science and politics of cancer, see Balshem (1993), Frank (1991), Sontag (1978), and Stacey (1997). Consider also these recent writings: Clorfene-Casten (1996) and Stabiner (1997). On prostate cancer, see Korda (1996).

6. On the history of hereditarian thinking about disease, see Haller (1971), Hutchinson (1884), Kevles (1985), Ludmerer (1972), and Rosenberg (1976). One exception to this trend in seeing Medelian genetics as peripheral to medical practice is Pernick (1996). See also Wailoo (1991).

The literature on the history and social implications of modern genetics is vast. See, for example, Judson (1979), Kay (1993), Keller (1992), Kevles and Hood (1992), Kitcher (1996), and Olby (1974). Among the therapeutic by-products envisioned is gene therapy. See Allen (1975), Culver (1996), Hubbard and Wald (1993), Lyon and Gorner (1995), Nelkin and Lindee (1995), Nichols (1988), and Wexler (1995).

7. Noted one *Wall Street Journal* reporter in 1992, "the vast majority of Americans support the use of gene-based therapy to treat disease, even though they don't know much about the emerging science, according to a new survey" (Tanouye, 1992; see also Purvis, 1992).

8. The subtitle to Brown's 1995 article on the infusion of capital into gene therapy research noted, however, that "Cautious observers note, however, that the fate of the new industry may hinge on a flurry of recently approved trials" (Brown, 1995) Moreover, the death of Jesse Gelsinger in a gene therapy trial for a disease other than CF has placed increasing scrutiny on this new industry—leading to the scaling back of hopes, new regulatory oversight, and the closing of some trials (Nelson and Weiss, 2000; Weiss, 2000; Weiss and Nelson, 1999, 2000).

REFERENCES

Allen, G. 1975. *Life Sciences in the Twentieth Century*. New York: Cambridge University Press.

Anderson, D. H. 1938. Cystic fibrosis of the pancreas and its relation to celiac disease: A clinical and pathological study. *American Journal of Diseases of Children* 56:344–399.

Angier, N. 1992. Panel permits use of genes in treating cystic fibrosis. *New York Times*, December 4, p. A28.

Angier, N. 1993a. Cystic fibrosis: Experiment hits a snag. *New York Times*, September 22, p. C12.

Angier, N. 1993b. Gene therapy begins for fatal lung disease: Cystic fibrosis patient inhales altered cold virus. *New York Times*, April 20, p. C5.

Baggot, B. X. 1998. Human gene therapy patents in the United States. *Human Gene Therapy* 9(January 1):151–157.

Balshem, M. 1993. *Cancer in the Community: Class and Medical Authority*. Washington, D.C.: Smithsonian Institution Press.

Barbero, G. J. 1972. Statement. P. 216 in National Heart, Blood Vessel, Lung and Blood Act of 1972: Hearings Before the Subcommittee on Public Health and Environment of the Committee on Interstate and Foreign Commerce, House of Representatives, 92nd U.S. Congress, H.R. 12571, 13715, 12460, 13500, S. 3323 to Amend the Public Health Service Act, April 25–26, 1972. Washington, D.C.: U.S. Government Printing Office.

Bates, B. 1992. *Bargaining for Life: A Social History of Tuberculosis*, 1976–1938. Philadelphia: University of Pennsylvania Press.

Bayer, R. 1987. *Homosexuality and American Psychiatry: The Politics of Diagnosis*. Princeton, N.J.: Princeton University Press.

Beaudry, P., M. Marks, D. McDougall, K. Desmond, and R. Rangel. 1980. Is anti-*pseudomonas* therapy warranted in acute respiratory exacerbations in children with cystic fibrosis? *Journal of Pediatrics* 97:144–147.

Beecher, H. 1959. *Experimentation in Man*. Springfield, Ill.: Thomas.

Beecher, H. 1966. Ethics and clinical research. *New England Journal of Medicine* 74:1354–1360.

Bliss, M. 1982. *The Discovery of Insulin*. Chicago: University of Chicago Press.

Boas, E. 1940. *The Unseen Plague: Chronic Illness*. New York: J. J. Augustin.

Booth, C. 1993. Clinical research. In W. Bynum and R. Porter, eds., *Companion Encyclopedia of the History of Medicine, Vol. 1*. New York: Routledge.

Boxerbaum, D., C. Doershuk, and L. Matthews. 1972. Use of antibictics [*sic*] in cystic fibrosis. *Journal of Pediatrics* 81:188.

Brady, D. 1990. Signals of hope: Gene therapy may cure cystic fibrosis. *Maclean's* 103(October 1):52.

Brandt, A. 1985. *No Magic Bullet: A Social History of Venereal Disease in the United States since 1880*. New York: Oxford University Press.

Brent, L. 1997. *A History of Transplantation Immunology*. New York: Academic Press.

Brown, K. 1995. Major pharmaceutical companies infuse needed capital into gene therapy research. *The Scientist* 9(November 13):1, 10.

Brumberg, J. 1988. *Fasting Girls: The Emergence of Anorexia Nervosa as a Modern Disease*. Cambridge, Mass.: Harvard University Press.

Bud, R. 1978. Strategy in American cancer research after World War II: A case study. *Social Studies of Science* 8:425–459.

Burns, M. W., and J. R. May. 1968. Bacterial precipitin in serum of patients with cystic fibrosis. *Lancet* 1:270–272.

Bynum, W. 1988. Reflections on the history of human experimentation. In S. Spicker et al., *The Use of Human Beings in Research: With Special Reference to Clinical Trials*. Boston, Mass.: Kluwer Academic Press.

Cantor, D. 1993. Cancer. In W. Bynum and R. Porter, eds., *Companion Encyclopedia of the History of Medicine, Vol. 1*. New York: Routledge.

Carey, J. 1993. The $600 million horse race. *Business Week*, August 23, p. 68.

Clorfene-Casten, L. 1996. *Breast Cancer: Poisons, Profits, and Prevention*. Monroe, Maine: Common Courage Press.

Cowley, G. 1993. Closing in on cystic fibrosis: Researchers are learning to replace a faulty gene. *Newsweek*, May 3, p. 56.

Culver, K. 1996. Gene Therapy: A Primer for Physicians. *Larchmont*, N.Y.: Mary Ann Liebert.

di Sant' Agnese, P., R. Darling, and G. Perera. 1953. Abnormal electrolyte composition of sweat in cystic fibrosis of the pancreas. *Pediatrics* 12:549–563.

DuValm, M. K. 1972. Testimony. P. 89 in National Heart, Blood Vessel, Lung, and Blood Act of 1972: Hearings Before the Subcommittee on Public Health and Environment of the Committee on Interstate and Foreign Commerce, House of Representatives, 92nd U.S. Congress, H.R. 12571, 13715, 12460, 13500, S. 3323 to Amend the Public Health Service Act, April 25–26, 1972. Washington, D.C.: U.S. Government Printing Office.

Endicott, K. M. 1957. The chemotherapy program. *Journal of the National Cancer Institute* 19:275–293.

Etheridge, E. 1992. *Sentinel for Health: A History of the Centers for Disease Control*. Berkeley: University of California Press.

Farber, S. 1944. Pancreatic function and disease in early life. V. Pathologic changes associated with pancreatic insufficiency in early life. *Archives of Pathology* 37:238–250.

Farber, S. 1945. Some organic digestive disturbances in early life. *Journal of the Michigan Medical Society* 44:587–594.

Farmer, P. 1992. *AIDS and Accusation: Haiti and the Geography of Blame*. Berkeley: University of California Press.

Farmer, P. 1999. *Infections and Inequalities: The Modern Plagues*. Berkeley: University of California Press.

Feudtner, C. 1995. The want of control: Ideas, innovations, and ideals in the modern management of diabetes mellitus. *Bulletin of the History of Medicine* 69:66–90.

Fox, D., and E. Fee, eds. 1990. *AIDS: The Burden of History*. Berkeley: University of California Press.

Fox, R., and J. Swazey. 1992. *Spare Parts: Organ Replacement in American Society*. New York: Oxford University Press.

Frank, A. 1991. *At the Will of the Body: Reflections on Illness*. Boston, Mass.: Houghton Mifflin.

Friedlander, W. J. 1992. The Bigelow-Simpson controversy: Still another early argument over the discovery of anesthesia. *Bulletin of the History of Medicine* 66(Winter):613–625.

Friedmann, T. 1994. The promise and overpromise of human gene therapy. *Gene Therapy* 1:217–218.

Fujimura, J. 1996. *Crafting Science: A Sociohistory of the Quest for the Genetics of Cancer*. Cambridge, Mass.: Harvard University Press.

Garrard, S., J. Richmond, and M. Hirsch. 1951. *Pseudomonas aeruginosa* infection as a complication of therapy in pancreatic fibrosis (Mucoviscidosis). *Pediatrics* 8(October):485.

Grade, D. 1999. In breast cancer data, hope, fear, and confusion. *New York Times*, January 26.

Hall, S. 1997. *A Commotion in the Blood: Life, Death, and the Immune System*. New York: Henry Holt.

Haller, J. 1971. *Outcasts from Evolution: Scientific Attitudes of Racial Inferiority, 1859–1900.* Urbana: University of Illinois Press.

Hamilton, J. O'C. 1993. A star drug is born. *Business Week*, August 23, pp. 66–68.

Harden, V. 1986. *Inventing the NIH: Federal Biomedical Research Policy, 1887–1937.* Baltimore, Md.: The Johns Hopkins University Press.

Harvey, A. M. 1981. *Science at the Bedside: Clinical Research in American Medicine, 1905–1945.* Baltimore, Md.: The Johns Hopkins University Press.

Hubbard, R., and E. Wald. 1993. *Exploding the Gene Myth: How Genetic Information is Produced and Manipulated by Scientists, Physicians, Employers, Insurance Companies, Educators, and Law Enforcers.* Boston, Mass.: Beacon Press.

Hutchinson, J. 1884. *The Pedigree of Disease.* London: Churchill.

Hyatt, A., B. Chipps, K. Kumor, E. Mellits, P. Lietman, and B. Rosenstein. 1981. A double-blind controlled trial of anti-*pseudomonas* chemotherapy of acute respiratory exacerbations in patients with cystic fibrosis. *Journal of Pediatrics* 99:307–311.

Isles, A., I. Maclusky, M. Corey, R. Gold, C. Prober, P. Fleming, and H. Levison. 1984. *Pseudomonas cepacia* infection in cystic fibrosis: An emerging problem. *Journal of Pediatrics* 104:206–210.

Jacyna, L. S. 1988. The laboratory and clinic: The impact of pathology on surgical diagnosis in the Glasgow Western Infirmary, 1875-1910. *Bulletin of the History of Medicine* 62:384–406.

Jones, J. 1981. *Bad Blood: The Tuskegee Syphilis Experiment.* New York: Free Press.

Judson, H. F. 1979. *The Eighth Day of Creation: Makers of the Revolution in Biology.* New York: Simon and Schuster.

Katz, J. 1972. *Experimentation with Human Beings.* New York: Russell Sage.

Kay, L. E. 1993. *The Molecular Vision of Life: Caltech, the Rockefeller Foundation, and the Rise of the New Biology.* New York: Oxford University Press.

Kearns, G. L., B. Hilman, and J. Wilson. 1982. Dosing implications of altered gentamicin disposition in patients with cystic fibrosis. *Journal of Pediatrics* 1000:312–318.

Keller, E. F. 1992. Nature, Nurture, and the Human Genome Project. In D. Kevles and L. Hood, eds., *The Code of Codes: Scientific and Social Issues in the Human Genome Project.* Cambridge, Mass.: Harvard University Press.

Kehvles, D. 1985. *In the Name of Eugenics: Genetics and the Uses of Human Heredity.* New York: Knopf.

Kevles, D., and L. Hood, eds. 1992. *The Code of Codes: Scientific and Social Issues in the Human Genome Project.* Cambridge, Mass.: Harvard University Press.

Kitcher, P. 1996. *The Lives to Come: The Genetic Revolution and Human Possibilities.* New York: Simon and Schuster.

Kolata, G. 1995. Gene therapy shows no benefit in two studies. *New York Times*, September 28, p. A24.

Korda, M. 1996. *Man to Man: Surviving Prostate Cancer.* New York: Random House.

Kulczycki, L., T. Murphy, and J. Bellanti. 1978. *Pseudomonas* colonization in cystic fibrosis: A study of 160 patients. *Journal of the American Medical Association* 240:30–34.

Leavitt, J. W. 1986. *Brought to Bed: Childbearing in America, 1750-1950.* New York: Oxford University Press.

Leavitt, J. 1996. *Typhoid Mary: Captive to the Public's Health.* Boston, Mass.: Beacon.

Lederer, S. 1995. *Subjected to Science: Human Experimentation in America Before the Second World War.* Baltimore, Md.: The Johns Hopkins University Press.

Loening-Baucke, V. A., E. Mischler, and M. G. Myers. 1979. A placebo-controlled trial of cephalexin therapy in the ambulatory management of patients with cystic fibrosis. *Journal of Pediatrics* 95:630–637.

Lowy, I. 1989. Biomedical research and the constraints of medical practice: James Bumgardner Murphy and the early discovery of the role of lymphocytes in immune reactions. *Bulletin of the History of Medicine* 63:356–391.

Lowy, I. 1996. *Between Bench and Bedside: Science, Healing, and Interleukin-2 in a Cancer Ward.* Cambridge, Mass.: Harvard University Press.

Ludmerer, K. 1972. *Genetics and American Society: A Historical Appraisal.* Baltimore, Md.: The Johns Hopkins University Press.

Lyon, J., and P. Gorner. 1995. *Altered Fates: Gene Therapy and the Retooling of Human Life.* New York: Norton.

Maier, P. 1972. The biggest public health experiment ever: The 1954 field trial of Salk polio vaccine. Pp. 2–13 in J. Tanur, ed., *Statistics: A Guide to the Unknown.* San Francisco: Holden-Day.

Mann, R. D. 1988. *The History of the Management of Pain: From Early Principles to Present Practice.* Park Ridge, N.J.: Parthenon.

Mann, C., and M. Plummer. 1991. *The Aspirin Wars: Money, Medicine, and One Hundred Years of Rampant Competition.* New York: Knopf.

Markle, G., and J. Petersen, eds. 1980a. *Politics, Science, and Cancer: The Laetrile Phenomenon.* Boulder, Colo.: Westview Press.

Markle, G., and J. Petersen. 1980b. The Laetrile Phenomenon: An overview. In G. Markle and J. Petersen, eds., *Politics, Science, and Cancer: The Laetrile Phenomenon.* Boulder, Colo.: Westview Press.

Marks, H. 1997. *The Progress of Experiment: Science and Therapeutic Reform in the United States, 1900–1990.* Cambridge, U.K.: Cambridge University Press.

Marshall, E. 1995. Gene therapies growing pains (the trouble with vectors). *Science* 269:1052.

Martin, E. 1994. *Flexible Bodies: Tracking Immunity in American Culture from the Days of Polio to the Age of AIDS.* Boston, Mass.: Beacon.

Martin, P., and S. Thomas. 1988. The commercial development of gene therapy in Europe and the USA. *Human Gene Therapy* 9(January 1):87–114.

Matthews, J. R. 1995. *Quantification and the Quest for Medical Certainty.* Princeton, N.J.: Princeton University Press.

McTavish, J. R. 1987. What's in a name? Aspirin and the American Medical Association. *Bulletin of the History of Medicine* 61:343–366.

Meldrum, M. 1994. *Departures from Design: The Randomized Clinical Trial in Historical Context, 1946–1970.* Ph.D. dissertation, State University of New York, Stony Brook.

Meltzer, A. 1990. Dr. Samuel J. Meltzer and intratracheal anesthesia. *Journal of Clinical Anesthesia* 2(January–February):54–58.

Morris, D. 1991. *The Culture of Pain.* Berkeley: University of California Press.

Moulin, A. 1989a. Immunology old and new: The beginning and the end. In P. Mazumdar, ed., *Immunology, 1930–1980: Essays on the History of Immunology.* Toronto: Wall and Thompson.

Moulin, A. 1989b. The immune system: A key concept in the history of immunology. *History and Philosophy of the Life Sciences* 11:13–28.

Nelkin, D., and M. S. Lindee. 1995. *The DNA Mystique: The Gene as a Cultural Icon.* New York: Freeman.

Nelson, D. A. 1993. Intraspinal therapy using methyprednisone acetate: Twenty-three years of clinical controversy. *Spine* 18(February):278–286.

Nelson, J. D. 1985. Discussion of Management of Acute Pulmonary Exacerbations in Cystic Fibrosis: A Critical Reappraisal. *Journal of Pediatrics* 104:206–210.

Nelson, D., and R. Weiss. 2000. FDA stops researcher's gene therapy experiments. *Washington Post*, March 2, p. A8.

New York Times. 1972. Poster child named. February 21, Section 2, p. 35.

New York Times. 1991. Drug by Genetech gets orphan status. January 30, p. D4.

New York Times. 1993. FDA approval sought for cystic fibrosis drug. March 31, p. D4.

Nichols, E. 1988. *Human Gene Therapy.* Cambridge, Mass.: Harvard University Press.

Nichols, M. 1993. A test case in hope. *Maclean's* (May 3):39.

Nolan, G., P. Moiror, H. Levison, P. Fleming, M. Corey, and R. Gold. 1982. Antibiotic prophylaxis in cystic fibrosis: Inhaled cephaloridine as an adjunct to oral cloxacillin. *Journal of Pediatrics* 101:626–630.

Ohlendorf, P. 1985. The taming of a once-certain killer. *Maclean's* 98(October 7):50, 52.

Olby, R. 1974. *The Path of the Double Helix.* Seattle: University of Washington Press.

Palca, J. 1994. The promise of a cure. *Discover,* June, p. 86.

Panem, S. 1984. *The Interferon Crusade.* Washington, D.C.: The Brookings Institute.

Papper, E. M. 1992. The influence of romantic literature on the medical understanding of pain and suffering—The stimulus to the discovery of anesthesia. *Perspectives in Biology and Medicine* 35(Spring):401–415.

Pappworth, M. H. 1967. *Human Guinea Pigs: Experimentation on Man.* London: Routledge.

Parry, M., H. Neu, M. Melino, P. Gaerlan, C. Ores, and C. Denning. 1977. Treatment of pulmonary infections in patients with cystic fibrosis: A comparative study of ticarcillin and gentamicin. *Journal of Pediatrics* 9:144–148.

Patterson, D. 1987. *The Dread Disease: Cancer and Modern American Culture.* Cambridge, Mass.: Harvard University Press.

Peitzmann, S. 1992. From Bright's disease to end-stage renal disease. In C. Rosenberg and J. Golden, eds., *Framing Disease: Studies in Cultural History.* New Brunswick, N.J.: Rutgers University Press.

Peller, S. 1979. *Cancer Research Since 1900: An Evaluation.* New York: Philosophical Library.

Pernick, M. 1985. *A Calculus of Suffering: Pain, Professionalism, and Anesthesia in Nineteenth-Century America.* New York: Columbia University Press.

Pernick, M. 1996. *The Black Stork: Eugenics and the Death of "Defective" Babies in American Medicine and Motion Pictures Since 1915.* New York: Oxford University Press.

Peters, T. 1993. Interferon and its first clinical trials: Looking behind the scenes. *Medical History* 37:270–295.

Pitcock, C. D., and R. B. Clark. 1992. From Fanny to Ferdinand: The development of consumerism in pain control during the birth process. *American Journal of Obstetrics and Gynecology* 157(September):581–587.

Plough, A. 1986. *Borrowed Time: Artificial Organs and the Politics of Extending Lives.* Philadelphia, Pa.: Temple University Press.

Proctor, R. 1995. *The Cancer Wars: How Politics Shapes What We Know and Don't Know about Cancer.* New York: Basic Books.

Purvis, A. 1992. Laying siege to a deadly gene. *Time,* February 24, p. 60.

Rather, L. J. 1978. *The Genesis of Cancer: A Study of the History of Ideas.* Baltimore, Md.: The Johns Hopkins University Press.

Rettig, R. 1977. *Cancer Crusade: The Story of the National Cancer Act of 1971.* Princeton, N.J.: Princeton University Press.

Rey, R. 1995. *The History of Pain.* Cambridge, Mass.: Harvard University Press.

Richards, E. 1991. *Vitamin C and Cancer: Medicine or Politics?* London: MacMillan.

Riordan, T., J. Rommens, B. Kerem, N. Alon, R. Rozmahel, Z. Grzekzak, J. Zelenski, S. Lok, N. Plavsic, J. L. Chou, M. Drumm, M. Ianuzzi, F. S. Collins, and L. Tsui. 1989. Identification of the cystic fibrosis gene: Cloning and characterization of complementary DNA. *Science* 245:1066–1073.

Rogers, N. 1992. *Dirt and Disease: Polio Before FDR.* New Brunswick, N.J.: Rutgers University Press.

Rosenberg, C. 1962. *The Cholera Years: The United States in 1832, 1849, and 1865.* Chicago: University of Chicago Press.

Rosenberg, C. 1976. The bitter fruit: Heredity, disease, and social thought. In C. Rosenberg, ed., *No Other Gods: On Science and American Social Thought.* Baltimore, Md.: The Johns Hopkins University Press.

Rosenberg, C., and J. Golden, eds. 1992. *Framing Disease: Studies in Cultural History.* New Brunswick, N.J.: Rutgers University Press.

Rosner, D., and G. Markowitz. 1991. *Deadly Dust: Silicosis and the Politics of Occupational Disease in Twentieth-Century America.* Princeton, N.J.: Princeton University Press.

Rothman, D. 1991. *Strangers at the Bedside: A History of How Law and Bioethics Transformed Medical Decision Making.* New York: Basic Books.

Rushman, G. B., N. J. H. Davies, and R. S. Atkinson. 1996. *A Short History of Anesthesia: The First Hundred and Fifty Years.* Oxford, U.K.: Butterworth-Heinemann.

Scarry, E. 1985a. *The Body in Pain: The Making and Unmaking of the World.* New York: Oxford University Press.

Scarry, E. 1985b. Willow-bark and red poppies: Advertising the remedies for physical pain. *Word and Image* 10:381–408.

Seligmann, J. 1990. Curing cystic fibrosis?: Genes convert sick cells. *Newsweek*, October 1, p. 64.

Seligmann, J., and D. Glick. 1989. Cystic fibrosis: Hunting down a killer gene. *Newsweek*, September 4, pp. 60–61.

Shulman, R. 1985. The gene doctors. *Business Week*, November 18, pp. 76–85.

Sigerist, H. 1932. The historical development of the pathology and therapy of cancer. *Bulletin of the New York Academy of Medicine* 8:642–653.

Smith, J. 1990. *Patenting the Sun: Polio and the Salk Vaccine.* New York: William Morrow and Co.

Sontag, S. 1978. *Illness as Metaphor.* New York: Penguin.

Stabiner, K. 1997. *To Dance with the Devil: The New War on Breast Cancer—Politics, Power, and People.* New York: Delta.

Stacey, J. 1997. *Teratologies: A Cultural Study of Cancer.* New York: Routledge.

Strickland, S. 1972. *Politics, Science, and Dread Disease: A Short History of United States Medical Research Policy.* Cambridge, Mass.: Harvard University Press.

Studer, K., and D. Chubin. 1980. *The Cancer Mission: Social Contexts of Biomedical Research.* London: Sage Publications.

Szasz, T. 1957. *Pain and Pleasure: A Study of Bodily Feeling.* London: Tavistock.

Tanouye, E. 1992. Majority supports gene-based therapy. *Wall Street Journal*, September 29, p. B10.

Tauber, A. 1994. *The Immune Self: Theory or Metaphor?* New York: Cambridge University Press.

Tweedie, I. E., and S. L. Snowdon. 1990. The Trilite inhaler: An historical review and performance assessment. *Anaesthesia* 45(September):757–759

Unruh, A. M. 1996. Review article: Gender variations in clinical pain experience. *Pain* 65:123–167.

Wailoo, K. 1991. "A disease sui generis:" The origins of sickle cell anemia and the emergence of modern clinical research. *Bulletin of the History of Medicine* 65:185–208.

Wailoo, K., ed. 1997a. *Drawing Blood: Technology and Disease Identity in Twentieth-Century America.* Baltimore, Md.: The Johns Hopkins University Press.

Wailoo, K. 1997b. The rise and fall of splenic anemia: Surgical identity and ownership of a blood disease. In K. Wailoo, ed., *Drawing Blood: Technology and Disease Identity in Twentieth-Century America.* Baltimore, Md.: The Johns Hopkins University Press.

Wailoo, K., and S. Pemberton. In press. *Genes, Risk, and Justice: The Transformation of Genetic Disease in America*. Baltimore, Md.: The Johns Hopkins University Press.

Wall, P. D. 1975. Editorial. *Pain* 1:1–2.

Warwick, W. J. 1977. Cystic fibrosis: An expanding challenge for internal medicine. *Journal of the American Medical Association* 238(November 14):2159–2162.

Wear, A. 1985. Historical and cultural aspects of pain. *Bulletin of the Society for the Social History of Medicine* 36:7–21.

Weiss, R. 2000. Caution over gene therapy puts hopes on hold. *Washington Post*, March 7, p. A1.

Weiss, R., and D. Nelson. 2000. FDA lists violations by gene therapy director at University of Pennsylvania. *Washington Post*, March 4, p. A4.

Weiss, R., and D. Nelson. 1999. Gene therapy's troubling crossroads: A death raises questions of ethics, profit, and science. *Washington Post*, December 31, p. A3.

Wexler, A. 1995. *Mapping Fate: A Memoir of Family Risk, and Genetic Research*. Berkeley: University of California Press.

Young, J. H. 1980. Laetrile in historical perspective. In G. Markle and J. Petersen, eds., *Politics, Science, and Cancer: The Laetrile Phenomenon*. Boulder, Colo.: Westview Press.

11

Genetics Meets Genomics: Trends and Implications

David Schlessinger

Throughout the twentieth century, geneticists have functioned in two worlds. One of them is the classical academic endeavor that identifies genes by looking at the inheritance of traits in peas and flies as well as humans. But geneticists have also been among those scientists consistently up front in a second world, that of public discourse.

The recipients of the James S. McDonnell Foundation Centennial Fellowships in human genetics epitomize the cutting edge in both of the geneticists' worlds. They also exemplify how both of these roles of geneticists have been changing decisively as the Human Genome Project has shifted into high gear.

THE HUMAN GENOME PROJECT
AND THE FIRST WORLD OF GENETICISTS

We have all become accustomed, in a surprisingly short time, to the notion that in the coming few years we will have at our disposal the sequence of all three billion nucleotides of human DNA, along with a complete catalog of genes for our own species as well as a series of model organisms. The assembly of these gene catalogs continues an historical transition in which genes continue to be inferred as genetic factors by observing inherited traits, but are also increasingly thought of as defined sequences of DNA. And that makes a very big difference. This is because

National Institute on Aging, Bethesda, Maryland

251

thinking about genetics based on populations and statistical studies requires special quantitative abilities and thereby sharply restricts the pool
of geneticists. But anyone can think about sequences of DNA.

Thus, as the catalog of genes is being completed and technology improves, almost all biologists are being recruited to the ranks of geneticists.
Some observers have recently suggested that medical schools may have
to be renamed genetic schools. In the analysis of genes, there has been a
concomitant sharp change as traditional studies, gene by gene, are replaced by systematic surveys across the entire hereditary potential of species. As mapped genes are identified along chromosomes and the corresponding proteins are being studied, there is a rapid transformation of
studies of biological function. Traditional genetics is merging with
genomics, the systematic study of genes as sequences that produce corresponding proteins. The result is an increasingly extensive functional
analysis (Figure 1, left panel and top right). The new "functional
genomics" comprises ambitious attempts to understand everything about
the genome, including systematic studies of gene expression in cell and
tissue physiology and of the consequences of mutational deficiency in
each gene.

In this combined genetic and genomic approach, the modern version
of the first world of geneticists, the analysis of genes proceeds hand in
hand with an increasing understanding of the structure and function of

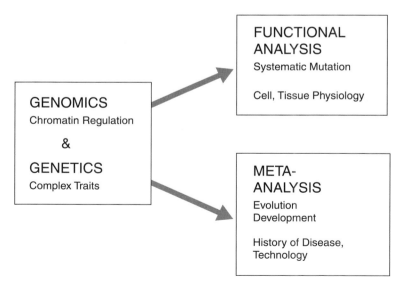

FIGURE 1   Combined genetic and genomic approach.

the cell nucleus where the genes are localized. In Wendy Bickmore's chapter, she has shown how extensions of fluorescent in situ hybridization can lead from a knowledge of cytogenetic changes in DNA to an analysis of chromosome dynamics during mitosis or chromatin remodeling. These are a vital feature, for example, of the famous cloning of the sheep Dolly. When the nucleus of a mature cell is reintroduced into an egg and is reprogrammed to reinitiate embryonic development, what happens to the chromosomes? Naturally, the cytogenetic analyses must be accompanied by extensive biochemical studies of the protein components and their interactions. But as in the past, we can expect cytogenetic hints to lead the way toward later biochemical and physiological investigations.

If we continue to think about genes as sequences of DNA, but associated with inherited "complex traits," we come to the discussion of the modern use of genomics in the Human Genome Project to provide a huge and increasing inventory of the repertoire of variation in DNA (Figure 1, left). In Leonid Kruglyak's chapter, this variation is the raw material that is required for the study of the genetic factor in disease causation. Variant sequences permit one to track down the genes that are participating in a process. Kruglyak provides a first approximation view of the approach that he is helping to create, designed to analyze complex traits with the use of these variant sequences or "markers" coming from the Human Genome Project. As he alludes to in his chapter, such genetic studies have already had many notable successes in explicating many single gene diseases. Although he emphasizes that the approach remains to be fully formulated and tested, I think that he subscribes to the general hope and optimism that the route he outlines will lead to a dramatic augmentation of the power to diagnose, anticipate, and perhaps alleviate complex chronic diseases.

META-ANALYSES AND THE SECOND WORLD OF GENETICISTS

In general, the first world of geneticists now thinks in terms of the convergence of genomics and genetics and the exploitation of the resultant functional analysis. But this view of trends, though it often dominates thinking in the field, is incomplete because, along with pure functional analysis, we now have a parallel enormous surge of meta-analysis of different types. These activities build on comparative studies of long-term changes in systems. An example is referred to by Leonid Kruglyak: the use of genetics and genomics to expand discussions of comparative evolution of all species, including our own. Similar extensions of developmental genomics and genetics are beginning to reveal the details of stages from fetal growth to aging and aging-associated conditions.

Among the most important meta-analyses are the ones that look reflectively at the way in which the techniques and changing models of genetics and genomics fit into ongoing transformations of our thinking about biology. This brings us to the second world of geneticists that I referred to above: the world that impacts on social policy and the treatment of patients. Of course the twentieth was indeed the century of Watson and Crick and all of us whose works are footnotes to theirs. One cannot ignore that the twentieth was also the century in which successive waves of immigrants to this country were labeled as genetically defective; in which eugenics movements were closely linked to fascist programs; and in which true geneticists in the Soviet Union were imprisoned or, worse, while Lysenko and his colleagues dictated the results that scientists were required to find in experiments.

A relatively small group of sociologists, ethicists, biological scientists, and historians have had the commitment and the broad base of knowledge to look critically at how heredity and disease are understood. And even fewer have the ability to use that knowledge to illuminate what we might do to ensure that the political and medical use of genetics is optimized for humanity. Keith Wailoo is one of those few contributors. His work and thinking operate at the level of meta-analysis of the exploration of twentieth century biomedicine, analyzing the way in which clinical science has looked at disease; how this has been influenced by new technology, and how the relationship of the science of disease and the experience of illness is changing.

Keith Wailoo's chapter requires no help from explication by others, but I reiterate that the interplay between the discussions of trends in genetics and trends in its effects on thinking about illness lies precisely at the intersection of the two worlds of geneticists: the analysis of function and its implications for public discussion and policy.

I believe that the research by Bickmore, Kruglyak, and Wailoo could be very useful and provocative in future discussions of genetics and its uses; and this is of course relevant to the purposes of the McDonnell Fellowship Program. I was one of those who enjoyed the privilege of having several discussions with Mr. McDonnell about genetics. He was quite consistent in his thinking about both the centrality and the power of genetics in modern life, and also about the necessity for scientists to combine scientific prowess with the communication skills to inform a free society. The chapters by Bickmore, Kruglyak, and Wailoo are appropriately cogent and also relatively free of the impenetrable jargon that affects most discussions of biology.

12

Scales that Matter: Untangling Complexity in Ecological Systems

Mercedes Pascual

En aquel Imperio, el Arte de la Cartografía logró tal Perfección que el Mapa de una sola Provincia ocupaba toda una Ciudad, y el Mapa del Imperio, toda una Provincia. Con el tiempo, esos Mapas Desmesurados no satisficieron y los Colegios de Cartógrafos levantaron un Mapa del Imperio, que tenía el tamaño del Imperio y coincidía puntualmente con él. Menos Adictas al Estudio de la Cartografía, las Generaciones Siguientes entendieron que ese dilatado Mapa era Inútil y no sin impiedad lo entregaron a las Inclemencias del Sol y de los Inviernos. . . .

> (Suárez Miranda, "Viajes de varones prudentes," Libro XIV, 1658)
> —Jorge Luis Borges y Adolfo Bioy Casares,
> "Del Rigor en la Ciencia" en "Cuentos Breves y Extraordinarios."

In that Empire, the Art of Cartography reached such Perfection that the Map of one Province alone took up the whole of a City, and the Map of the Empire, the whole of a Province. In time, those Unconscionable Maps did not satisfy and the Colleges of Cartographers set up a Map of the Empire which had the size of the Empire itself and coincided with it point by point. Less Addicted to the Study of Cartography, Succeeding Generations understood that this widespread Map was Useless and not without Impiety they abandoned it to the Inclemencies of the Sun and of the Winters . . . .

> —Translation from Jorge Luis Borges (1964).

Center of Marine Biotechnology, University of Maryland Biotechnology Institute

ABSTRACT

Complexity in ecological systems results not only from a large number of components, but also from nonlinear interactions among the multiple parts. The combined effects of high dimensionality and nonlinearity lead to fundamental challenges in our ability to model, understand, and predict the spatiotemporal dynamics of ecological systems. In this chapter I argue that these challenges are essentially problems of scale—problems arising from the interplay of variability across scales. I address two main consequences of such interplay and sketch avenues for tackling related problems of scale with approaches at the interface of dynamical system theory and time-series analysis.

The first consequence involves a "scale mismatch" between fluctuations of the physical environment and those of ecological variables: In a nonlinear system, forcing at one scale can produce an ecological response with variability at one or more different scales. This scale mismatch challenges our ability to identify key environmental forcings that are responsible for ecological patterns with conventional statistical approaches based on assumptions of linearity.

In the first half of this chapter, theoretical models are used to illustrate the rich array of possible responses to forcing. The models focus on systems for antagonistic interactions, such as those for consumers and their prey or pathogens and their hosts. These examples motivate an important empirical question: How do we identify environmental forcings from ecological patterns without assuming a priori that systems are linear? An alternative approach is proposed, based on novel time-series methods for nonlinear systems. Its general framework should also prove useful to predict ecological responses as a function of environmental forcings. General areas for future application of this approach are outlined in two fields of primary importance to humans—fisheries and epidemiology. In these fields, the role of physical forcings has become the subject of renewed attention, as concern develops for the consequences of human-induced changes in the environment.

The nonlinear models used in the proposed approach are largely phenomenological, allowing for unknown variables and unspecified functional forms. As such, they are best suited for systems whose ecological interactions are well captured by a low number of variables. This brings us to the second part of this chapter, the relationship between scale of description and dynamical properties, including dimensionality.

Here, a second consequence of the interplay of variability across scales is considered, which involves the level of aggregation at which to sample or model ecological systems. This interplay complicates in fundamental ways the problem of aggregation by rendering simple averaging impossible. Aggregation is, however, at the heart of modeling ecological systems—of defining relevant variables to represent their dynamics and of simplifying elaborate models whose high dimensionality precludes understanding and robust prediction.

The second part of this chapter addresses the specific problem of selecting a spatial scale for averaging complex systems. A spatial predator–prey model following the fate of each individual is used to illustrate that fundamental dynamical properties, such as dimensionality and determinism, vary with the spatial scale of averaging. Methods at the interface of dynamical system theory and time-series analysis prove useful to describe this variation and to select, based on it, a spatial scale for averaging. But scale selection is only the first step in model simplification. The problem then shifts to deriving an approximation for the dynamics at the selected scale. The dynamics of the predator–prey system serve to underscore the difficulties that arise when fine-scale spatial structure influences patterns at coarser scales. Open areas for future research are outlined.

Ecologists are well aware of the complicated dynamics that simple, low-dimensional systems can exhibit even in the absence of external forcings. The present challenge is to better understand the dynamics of high-dimensional nonlinear systems and their rich interplay with environmental fluctuations. Here, the perplexing order that nonlinearity creates opens doors to reduce complexity by exploiting relationships between dynamical properties and scale of description. This simplification is important for understanding and predicting dynamics and, ultimately, for better managing human interactions with the environment.

INTRODUCTION

Borges' "Map of the Empire" gives perhaps a lucid image of attempts to comprehend the dynamics of ecological systems by adding increasing knowledge on their parts. The complexity of such systems leads to fundamental challenges in understanding and predicting their spatiotemporal patterns. Yet ecological systems are essentially dynamic, changing continuously in space and time, and the understanding of this variation is crucial to better understand and manage human interactions with the environment.

Complexity is generally associated with the multiplicity of parts. As such, it is found in a variety of ecological systems. In food webs, for example, the number of trophic categories is large and can rapidly multiply as the observer describes trophic links in greater detail (Abrams et al., 1996). In studies of ecological communities, the large number of species has generated an active debate on the relevant variables and the level of organization at which biodiversity should be defined (Levin and Peale,

1996). The debate is clearly illustrated in the current research on the aggregation of species into functional groups (Dawson and Chapin, 1993; Deutschman, 1996; Rastetter and Shaver, 1995; Solbrig, 1993; Steneck and Dethier, 1994), or in the provocative title "Species as noise in community ecology: Do seaweeds block our view of the kelp forest" (Hay, 1994).

Beyond high dimensionality, however, complex systems possess another defining ingredient, *nonlinearity*, without which multiple interacting variables would not pose such a formidable challenge. Many ecological rates, such as those associated with per capita population growth, predation, and competition for resources, are not constant but vary as a function of densities. This density dependence in intraspecific and interspecific interactions leads to nonlinearity in ecological systems and in the models we build to capture their dynamics.

Nonlinearity creates a rich array of possible dynamics. Of these, chaos has occupied center stage. First encountered at the turn of this century (Dunhem, 1906; Hadamard, 1898; Poincaré, 1908), chaos raised the unexpected and somewhat disturbing possibility[1] of aperiodic dynamics with sensitivity to initial conditions in deterministic systems. Many decades passed before its definite discovery in the work of Lorenz (1963), Ruelle and Takens (1971), May (1974), and others. A watershed followed in nonlinear dynamics research. Hassell et al. (1976) and Schaffer and Kot (1985, 1986) first applied approaches from nonlinear dynamical systems to ecological data; and the debate continues on the relevance of chaos to natural systems (e.g., Ellner and Turchin, 1995; Hastings et al., 1993; Sugihara, 1994). Although chaos remains a fascinating concept, it is on a different and more ubiquitous property of nonlinearity that I wish to focus my attention here, a property of fundamental importance to understanding, predicting, and modeling the dynamics of complex ecological systems.

In nonlinear dynamical systems, variability interacts *across* scales in space and/or time with consequences far beyond those emphasized so far by ecological studies centered on chaos. Indeed, a main message arising from such studies has been that complicated dynamics are possible in simple, low-dimensional systems, and that highly irregular fluctuations in population numbers can occur in the absence of forcing by environmental fluctuations. Here, I shift the focus to the high-dimensional nature of ecological systems and to their rich interplay with fluctuations of the physical environment.

But how does the interplay of variability *across* scales matter to ecology? One important consequence of this interplay involves what I call a "scale mismatch" between the fluctuations of the physical environment and those of ecological variables: In a nonlinear system, forcing at one scale can produce an ecological response with variability at one or more *different* scales. This scale mismatch challenges our ability to identify key environmental forcings that are responsible for ecological patterns. Indeed, a common approach, implemented with well-known methods such as cross-correlation and cross-spectral analysis, relies on matching scales of variability. This approach concludes that an ecological pattern is caused by a physical factor if their variances share a dominant period or wavelength. Denman and Powell (1984) give numerous examples of successful results with this approach in a review of plankton patterns and physical processes. They point out, however, that as often as not, ecological responses could not be linked to a particular physical scale. One possible explanation is nonlinearity: scale-matching methods should be most successful at establishing cause and effect relationships in linear systems, or close to equilibria, where nonlinear systems are well approximated by linear ones. There is, however, ample evidence for nonlinearity in population growth, ecological interactions, and the response of ecosystems to perturbations (Denman and Powell, 1984; Dwyer and Perez, 1983; Dwyer et al., 1978; Ellner and Turchin, 1995; Turchin and Taylor, 1992).

Another important consequence involves the scale at which to sample or model the dynamics of complex systems and the related problem of aggregating their components in space, time, or level of organization. In nonlinear systems, the interplay of variability across scales complicates in fundamental ways the problem of aggregation by rendering simple averaging impossible (Steele et al., 1993). Aggregation is, however, at the heart of modeling ecological systems—of defining relevant variables to represent and understand their dynamics. Examples include the definition of functional groups in communities (Hay, 1994), trophic levels in food webs (Armstrong, 1994), demographic classes in populations (Caswell and John, 1992), and the density of individuals in spatial systems (Pascual and Levin, 1999b).

To represent the complexities of distributed interactions in fluctuating environments, realistic ecological models increasingly incorporate elements of space, multiple interactions, and stochasticity. However, in the high dimensionality of these models lies both their strength and their weakness: strength because the systems they represent are undoubtedly high dimensional and weakness because one must generally study them by performing large simulations. It is therefore difficult to elucidate the essential processes determining their spatial and temporal dynamics and the critical mechanistic bases of responses to changing environmental conditions. One route to establish such a link is to simplify models to their essentials by aggregating their basic units.

In spatial systems, simplification involves deriving approximations for the dynamics at a coarser spatial scale (Levin and Pacala, 1997). Beyond simplification, this process of scaling dynamics addresses the importance of local detail (Pacala and Deutschman, 1996), the identification of the scale at which macroscopic descriptions become effectively deterministic (Keeling et al., 1997; Pascual and Levin, 1999b; Rand and Wilson, 1995), and how to represent variability at scales smaller than those that are explicitly included in a model (Steele et al., 1993). These issues pervade the study of complex systems but are especially relevant to ecology where many and perhaps most measurements are taken at smaller scales than those of the patterns of interest.

In summary, problems of scale set fundamental limits to the understanding and prediction of dynamics in complex ecological systems. Below I address the two specific problems of identifying key environmental forcings and selecting a spatial scale for averaging. These are treated, respectively, in the first and second half of the chapter. I propose that methods at the interface of dynamical systems theory and time-series analysis have important applications to problems of scale. Throughout the chapter, connections are drawn to marine systems.

Although biological oceanography has pioneered the concept of scale in ecology (Haury et al., 1978; Steele, 1978), a linear perspective has dominated the view of how environmental and biological scales interact. A few authors have cautioned against simple linearity assumptions (Denman and Powell, 1984; Dwyer and Perez, 1983; Star and Cullen, 1981; Steele, 1988). Here, I further support the need for a nonlinear perspective.

NONLINEAR ECOLOGICAL SYSTEMS AND FLUCTUATING ENVIRONMENTS

The relations between the two of them must have been fascinating. For things are not what they seem.

—Saul Bellow, *It All Adds Up:*
From the Dim Past to the Uncertain Future

There is in ecology a long history of dispute between explanations of population and community patterns based on intrinsic biological processes (competition, predator–prey interactions, density dependence, etc.) and explanations based on extrinsic environmental factors (spatial gradients and/or temporal changes in temperature, moisture, light intensity, advection, etc.). In population ecology, Andrewartha and Birch (1954) considered density-independent factors, such as weather, the primary determinants of population abundance. In this view, chance fluctuations played an important role (Birch, 1958). By contrast, Smith (1961) and Nicholson (1958) viewed populations as primarily regulated by density-dependent factors, such as competition, with densities continually tending toward a state of balance in spite of environmental fluctuations (Kingsland, 1985). In community ecology, the concept of interacting species as an integrated whole—a superorganism—in which competition leads to a climax stable state (Clements, 1936) contrasted with that of a loose assemblage dominated by stochastic fluctuations (Gleason, 1926). An undercurrent of this debate reemerged with the discovery of chaotic dynamics in the form of the dichotomy "noise versus determinism," and it is still detectable today [see, for example, the discussions on the power spectra of population time series (Cohen, 1996; Sugihara, 1996)].

Mathematical approaches to ecological theory were strongly influenced by Clements' community climax and Nicholson's population balance (Kingsland, 1985). These ideas found a natural representation in the mathematical concepts of equilibrium and local stability, becoming amenable to analysis within the realm of linear systems. In fact, nonlinear models for density-dependent interactions are well approximated by linear equations, as long as the system lies sufficiently close to equilibrium.

In the past two decades, the concepts of equilibrium and local stability have slowly faded away, with the growing recognition of the dynamic nature of communities and of the full array of possible dynamics in nonlinear models (e.g., May, 1974; Paine and Levin, 1981; Pickett and White, 1985; Schaffer and Kot, 1985). Today, studies of nonlinear systems are also moving away from the simple dichotomy "determinism versus noise" to encompass the interplay of nonlinear ecological interactions with stochasticity (e.g., Ellner and Turchin, 1995; Pascual and Levin, 1999b; Sugihara, 1994). Surprisingly, however, a linear perspective continues to dominate statistical approaches for relating ecological patterns to environmental fluctuations.

I illustrate below, with an array of nonlinear ecological models, the rich potential interplay of intrinsic (biological) processes and the external environment. This interplay limits the ability of conventional time-series methods, such as cross-correlation analysis, to detect environmental forcings that are responsible for population patterns. I outline an alternative approach based on novel time-series methods for nonlinear systems. I end this first half of the chapter with general areas for future application of this approach—fisheries and epidemiology. In these two fields of primary importance to humans, the role of the physical environment has become the subject of renewed attention as concern develops for the consequences of human-induced changes in the environment, particularly in climate.

A Scale Mismatch: Some Theoretical Examples

Ecological models can identify specific conditions leading to scales of variability in population patterns that are different from those in the underlying environment. The following examples center around one common nonlinear ecological interaction, that between consumers and their resources. The antagonistic interaction between a predator and its prey is well known for its unstable nature leading to oscillatory behavior in the form of persistent cycles or transient fluctuations with slow damping. Such cycles introduce an intrinsic temporal frequency that is capable of interacting with environmental fluctuations.

One well-known example of this interaction is given by predator–prey models under periodic forcing, where one parameter, such as the growth rate of the prey, varies seasonally. These models display a rich array of possible dynamics, including frequency locking, quasi-periodicity, and chaos (Inoue and Kamifukumoto, 1984; Kot et al., 1992; Rinaldi et al., 1993; Schaffer, 1988). In these dynamic regimes, predator and prey can display variability at frequencies other than that of the seasonal forcing. For example, chaotic solutions are aperiodic and exhibit variance at all frequencies.

These predator–prey models have shown that a temporal-scale mismatch can occur when an endogenous and an exogenous cycle interact. Recent work has extended this result to another type of endogenous oscillation, known as generation cycles, which are profoundly different from the prey-escape cycles in predator–prey systems (Pascual and Caswell, 1997a). Generation cycles occur because of interactions between density-dependent processes and population structure—for instance, the population heterogeneity that is due to individual variation in age, size, or developmental stage (deRoos et al., 1992; Nisbet and Gurney, 1985; Pascual and Caswell, 1997a). This individual variation is not accounted for in typical consumer resource models with aggregated variables such as total biomass or density. It introduces high dimensionality in the form of multiple demographic classes or distributed systems (e.g., Tuljapurkar and Caswell, 1996).

Population heterogeneity can play an important role in consumer resource dynamics in variable environments. This was demonstrated with models for the dynamics of a phytoplankton population and a limiting nutrient resource in the experimental system known as the chemostat (Pascual and Caswell, 1997a). The chemostat provides a simple, yet controllable, idealization of an aquatic system with both an inflow and an outflow of nutrients. Most models for phytoplankton ignore population structure and group all cells in a single variable such as total biomass or density. However, a cell does have a life history, the cell division cycle. Each cell progresses through a determinate sequence of events preceding cell division, and the population is distributed in stages of the cell cycle (Figure 1). Models that include population structure (the stages of the cell cycle) can generate oscillatory dynamics under a *constant* nutrient supply, a behavior that is absent in the corresponding unstructured models

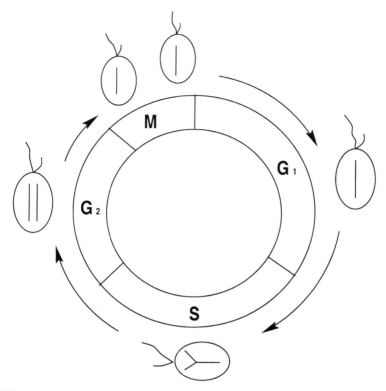

FIGURE 1 The cell cycle. The cell cycle is classically divided into four stages. The cell genome replicates during S, M corresponds to mitosis and cell division, and G1 and G2 denote stages during which most of cell growth takes place. SOURCE: Pascual and Caswell (1997a).

(Pascual, 1994). When forced by a periodic nutrient supply, the models with population structure exhibit aperiodic behavior with variability at temporal scales different from that of the forcing (Figure 2). As a result, the cross correlation between population numbers and nutrient forcing is low for any time lag, and observations of such a system would suggest only a weak link between phytoplankton patterns and nutrient input (Pascual and Caswell, 1997a).

Beyond population structure, another source of high dimensionality in ecological systems is the spatial dimension. Space can fundamentally alter the dynamics of consumer resource systems, as shown with a variety of reaction-diffusion models for interacting populations and random movement (see Okubo, 1980, or Murray, 1989, for a general review). This

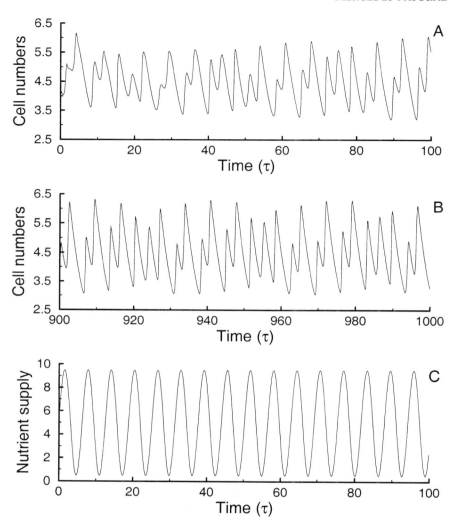

FIGURE 2   Population response to a periodic nutrient forcing. The forcing in the model is the periodic nutrient supply shown in (C). The population response is shown in (A) and (B) for initial and long-term dynamics, respectively. The dynamics of total cell numbers is quasi-periodic and displays variability at frequencies other than that of the forcing. SOURCE: Pascual and Caswell (1997a).

type of model has been used in a biological oceanography to model planktonic systems in turbulent flows. Theoretical studies have focused primarily on the formation of biological patterns in homogeneous environments (Levin and Segel, 1976, 1985). In heterogeneous environments, a biological pattern also emerges, which can differ strongly from the underlying spatial forcing (Pascual and Caswell, 1997b). This was shown with a model for a predator and a prey that interact and diffuse along an environmental gradient. Figure 3 shows the resulting irregular spatiotemporal patterns of prey numbers. Weak diffusion on a spatial gradient drives the otherwise periodic predator–prey model into quasi-periodic or chaotic dynamics (Pascual, 1993). In these regimes, the spatial distributions of predator and prey differ from the underlying environmental gradient, with a substantial lack of correlation between population and environmental patterns (Pascual and Caswell, 1997b).

The conditions for a scale mismatch identified by consumer resource models do exist in nature. For example, in the plankton there is evidence for the propensity of predator–prey interactions to oscillate both in the field (McCauley and Murdoch, 1987) and in laboratory experiments (Goulden and Hornig, 1980). There is also evidence for the occurrence of zooplankton generation cycles in the field (deRoos et al., 1992) and for phytoplankton oscillations in chemostat experiments, both transient and persistent, but not accounted for by traditional models without population structure (Caperon, 1969; Droop, 1966; Pickett, 1975; Williams, 1971). The cell cycle provides an explanation for the time delays previously invoked to explain these oscillatory transients (Pascual and Caswell, 1997a).

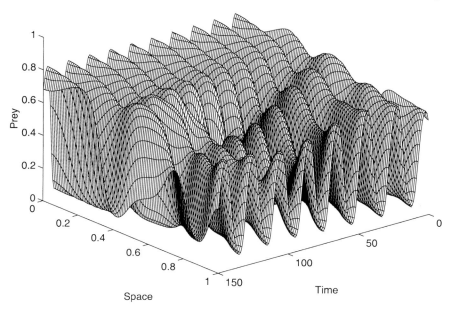

FIGURE 3 Irregular dynamics of prey numbers result from the weak diffusion of prey and predator along an environmental gradient. SOURCE: Pascual (1993).

Above the population level, Dwyer et al. (1978) and Dwyer and Perez (1983) provide compelling evidence for a scale mismatch in plankton dynamics. Their analysis of a 15-year time series showed that phytoplankton abundance in a temperate marine ecosystem displayed variability at multiple harmonics of the seasonal forcing frequency (Dwyer et al., 1978). A series of microcosm experiments, designed to further examine this response, exhibited plankton variability at a number of frequencies, none of which coincided with that of the sinusoidal forcing (Dwyer and Perez, 1983).

I end this series of examples by briefly touching on nonlinear systems with noise. Stochasticity, the extreme manifestation of high dimensionality represents in ecological models not only the unpredictable fluctuations of the environment but also the uncertainty arising from low population numbers. Recent results on the dynamics of epidemiological models give us a glimpse of the unexpected consequences of stochasticity in nonlinear systems, an area in which surprises are likely to continue and multiply. These models have revealed an elaborate interplay of endogenous cycles with periodic and *stochastic* forcings (e.g., Engbert and Drepper, 1994; Rand and Wilson, 1991; Sidorowich, 1992). In the models, one parameter, the contact rate, varies seasonally. In the absence of any forcing, the attractor of the system is a limit cycle: The long-term solutions are periodic. With a seasonal contact rate but no stochasticity, the limit cycle can coexist in phase space with a fascinating structure known as a repellor. Repellors represent the unstable counterparts of the more familiar strange attractors, that is, the geometrical objects in phase space onto which chaotic solutions relax as transients die out. Solutions are attracted toward the stable limit cycle but are continuously pushed away, against the unstable repellor, by the stochastic forcings (Rand and Wilson, 1991). In this way, solutions continuously switch between short-term periodic episodes, determined by the limit cycle, and chaotic transients, revealing the shadow of an unstable invariant set. These transients can be long lasting because trajectories take a long time to escape from the vicinity of the repellor. In addition, because the repellor influences these transients, solutions appear irregular and exhibit variability at a variety of scales different from the periodic forcing.

Beyond Linear Methods

The above examples raise an important empirical question: How can we identify environmental forcings related to specific population patterns when scales do not match? Furthermore, how do we approach this problem when all the relevant variables are not known and when the time series are short and noisy, two common properties of ecological data?

Recent methods at the interface of dynamical systems and time-series analysis may provide an answer.

I should pause to admit that a scale mismatch will not necessarily occur or be pronounced in *all* nonlinear systems. Even so, conventional methods might fail to detect relationships between ecological and environmental variables simply because their underlying assumption of linearity is violated. Thus, our main question can be restated in a more general form: How do we identify environmental forcings that are related to specific ecological patterns without assuming a priori that systems are linear?

The basic approach consists of modeling the dynamics of a variable of interest, $N(t)$, such as population density, with a nonlinear equation of the form

$$N_{t+Tp} = f[N_t, N_{t-\tau}, N_{t-2\tau}, \ldots, N_{t-(d-1)\tau}, E_{t-\tau_f}] \tag{1}$$

or

$$\frac{dN}{dt} = f[N_t, N_{t-\tau}, N_{t-2\tau}, \ldots, N_{t-(d-1)\tau}, E_{t-\tau_f}] \tag{2}$$

where Tp is a prediction time, f is a nonlinear function, τ is a chosen lag, d is the number of time-delay coordinates, and $E_{t-\tau_f}$ represents an exogenous variable lagged by τ_f (see below). This basic equation has three key features.

First, the function f is not specified in a rigid form. Instead, the functional form of the model is determined by the data (technically, the model is nonparametric).[2] This is appealing because we often lack the information to specify exact functional forms.

Second, the model uses time-delay coordinates. This is rooted in a fundamental result from dynamical systems theory, known as "attractor reconstruction" (Takens, 1981), which tackles the problem of not knowing (and therefore not having measured) all the interacting variables of a system. Takens' theorem essentially tells us that we can use time-delay coordinates as surrogates for the unobserved variables of a system (Figure 4). Although the diagram in Figure 4 considers a simple type of dynamics, the approach extends more generally to complex dynamics. Specifically, if the attractor of the system lies in an n-dimensional space, but one only samples the dynamics of a single variable x_t, then, for almost every time lag τ and for large enough d, the attractor of the d-dimensional time series

$$X_t = [x_t, x_{t-\tau}, x_{t-2\tau}, \ldots, x_{t-(d-1)\tau}] \tag{3}$$

is qualitatively similar to the unknown attractor of the n-dimensional system (Takens, 1981; for an ecological discussion, see Kot et al., 1988). The "embedding dimension" d, which needs to be sufficiently high but not larger than $2n+1$, corresponds to the notion of degrees of freedom, in the sense of providing a sufficient number of variables to specify a point on the attractor (Farmer, 1982).

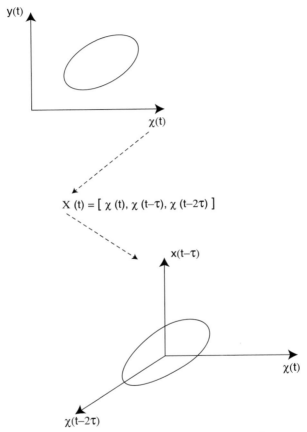

FIGURE 4   Sketch of "attractor reconstruction." Suppose the system of interest has two variables, $x(t)$ and $y(t)$. If both variables are measured, we can visualize the attractor of the system in phase space by plotting $y(t)$ versus $x(t)$ as a function of time (A). For sufficiently large time, once the effects of initial conditions die out, the resulting trajectory corresponds to the attractor of the system. For periodic dynamics, this attractor is a closed curve known as a limit cycle and is shown in (A). Now suppose that a single variable, $x(t)$, is measured. Then, to "reconstruct" the attractor, a multidimensional time-series is produced in which each coordinate is a lagged value of $x(t)$ [(B), where τ is a fixed lag]. The phase diagram obtained by plotting $x(t)$ versus $x(t - \tau)$ versus $x(t - 2\tau)$ exhibits an attractor that is topologically equivalent to that of the original system (C).

Third, the model incorporates the effect of an exogenous variable E_t, which influences the dynamics without being altered itself by the state of the system.[3] When stochastic, exogenous variables are known as dynamic noise; when deterministic, they can represent trends or periodic patterns. Examples include disturbances such as mortality due to storms, seasonal changes in parameters such as temperature or mixed layer depth, and interannual fluctuations in climate parameters such as those associated with the El Niño Southern Oscillation. The inclusion of an exogenous variable E_t in Equations (1) or (2) is based on an extension by Casdagli (1992) of Takens' theorem to input–output systems.

Nonlinear (nonparametric) time-series models, such as those in Equations (1) or (2), have been applied recently in ecology to questions on the qualitative type of dynamics of a system and on sensitivity to initial conditions (e.g., Ellner and Turchin, 1995). Beyond these questions, the models have potential but largely unexplored applicability to ecological problems involving environmental variability. Specific applications include (1) identifying the correct frequency of a periodic forcing by distinguishing among different candidate frequencies, (2) identifying environmental forcing(s) by distinguishing among candidate variables for which time series are available, (3) comparing models with and without environmental inputs, and (4) predicting the dynamics of an ecological variable as a function of environmental forcings. Future studies should carefully examine the criteria for model selection, as related not only to accuracy (how well the model fits the data) but also to predictability.

The proposed general framework will apply to systems whose internal dynamics are well captured by a finite and low number of variables (the embedding dimension d). This does not necessarily preclude its application to high-dimensional systems. The attractors of many infinite dimensional dynamical systems are indeed finite dimensional (Farmer, 1982). Moreover, as illustrated in the second part of this chapter, attractor dimension in stochastic nonlinear systems is a function of the scale of description.

A few words of caution. As for correlations, the phenomenological models described here do not necessarily imply causal relationships among variables. They can, however, contribute to the uncovering of such relationships and complement more mechanistic investigations. Here I have argued that incorporating specific information on environmental forcings is a useful yet largely unexplored avenue.

A Glimpse at Future Applications

Two areas of ecological research of primary concern to humans—fisheries and epidemiology—appear as good candidates for examining the usefulness of the above methods. Although seemingly unrelated, they do share some important prerequisites: the availability of data, the presence of nonlinearity in standard dynamical models, and the current interest in environmental effects on their dynamics.

The field of human epidemiology presents some of the best and longest data sets available for ecological research. Not surprisingly, this field has already seen the rich interplay of data analysis and dynamical modeling (e.g., Bolker and Grenfell, 1996; Grenfell and Dobson, 1995; Mollison, 1995; Olsen et al., 1988). Models for the dynamics of human disease are typically variants of a system of differential equations known as SEIR (for the fractions of susceptible, exposed, infectious, and recovered individuals in the host population) (Anderson and May, 1991). SEIR models are nonlinear and display a range of possible dynamics, from limit cycles to chaos (Kot et al., 1988; Schaffer et al., 1990).

For many diseases, the role of the physical environment has become an important current issue (Patz et al., 1996; World Health Organization, 1990), albeit a controversial one in light of climate change and certain speculative predictions (Taubes, 1997). Independent of climate change, however, better knowledge is required on the effects of climate *variability*. This knowledge can contribute to the implementation of timely preventive measures, particularly in regions of the globe where resources for prevention are either limited or not in place (Bouma and van der Kaay, 1994; Bouma et al., 1994). Climate variability has important manifestations in phenomena such as El Niño and in patterns of rainfall and temperature, which can influence disease dynamics (Bouma et al., 1994; Epstein et al., 1993). An ecological perspective on the dynamics of disease is not completely new. Already present in the first part of the century (Tromp, 1963), this perspective later lost strength, perhaps because medical developments allowed successful intervention, but also because advances in biological research opened the doors to more reductionistic investigations of pathogens and pathogen–host interactions. Today, the increased resistance of pathogens, as well as the breakdown of public health measures in some regions of the globe, argue for complementary research at the environmental level.

Recent studies of one particular human disease related to aquatic environments, cholera, exemplify the growing ecological perspective in infectious disease research (Colwell, 1996). These studies have shown that *Vibrio cholerae*, the causative agent of epidemic cholera, is a member of the autochthonous microbial flora of brackish water and estuaries (Colwell et al., 1981). Furthermore, experimental work has revealed a relationship between the survival of the bacterium and its attachment to plankton, suggesting a biological reservoir for *V. cholerae* in nature (Epstein et al., 1993; Huq et al., 1983). The importance of the reservoir and related environmental conditions to the patterns of cholera outbreaks remains an area of active research.

The effects of the physical environment are also the subject of renewed attention in fisheries. The International Council for the Exploration of the Sea recently held a symposium on physical–biological interactions in the recruitment dynamics of exploited marine populations in which a major section was devoted to climate variability and recruitment processes. A paper in the recent Special Report by Science on Human-Dominated Eco-systems calls for achieving "better holism" in fisheries management by consideration of broad-scale *physical forcing* and multiple species (Botsford et al., 1997). As for epidemiology, nonlinearity is present in the standard dynamical models of fisheries, known as stock-recruitment maps (Cushing, 1983). In spite of considerable efforts, prediction with these models has remained a difficult if not elusive goal (Rothschild, 1986).

THE SCALING OF COMPLEX SYSTEMS

> If this is how things stand, the model of models Mr. Palomar dreams of must serve to achieve transparent models, diaphanous, fine as cobwebs, or perhaps even to dissolve models . . . .

—Italo Calvino, *Palomar*[4]

Phenomenological models, such as the ones outlined in the first part of this chapter, provide a flexible framework to explore dynamic interactions and to attempt predictions. Ultimately, to better understand the dynamics of a complex system, more mechanistic models must be developed that incorporate the details of known variables and distributed interactions.

In population and community ecology, the extreme implementation of detail is found in models following the fate of each individual. Individual-based models are becoming increasingly popular, in part because technical developments allow us to perform large computer simulations and to sample at increasing resolutions. More importantly, individuals are both a fundamental unit of ecological interaction and a natural scale at which to make measurements, and individual-based models provide a mechanistic foundation that promises a sounder basis for understanding than do purely phenomenological models (Huston et al., 1988; Judson, 1994). Because of their high dimensionality, however, they are extremely sensitive to parameter estimation and prone to error propagation; the potential for analysis is limited, replaced in part by extensive and large simulations (Levin, 1992). It is therefore difficult to understand dynamics and to elucidate critical processes. To address these deficiencies, one is led naturally to issues of aggregation and simplification: When is variability at the individual level essential to the dynamics of densities? At what spatial scale should densities be defined? How do we derive equations for densities in terms of individual behavior? How does predictability vary as a function of spatial scale? (For examples of recent studies in this area, see Bolker and Pacala, 1997; Levin and Durrett, 1996; Pacala and Deutschman, 1996; Pascual and Levin, 1999a, 1999b). These questions are not limited to the aggregation of individuals, but apply more generally to any basic units on which the local interactions of a model are defined. Indeed, spatial aggregation is also an important problem in landscape and ecosystem models (King, 1992; Levin, 1992; Rastetter et al., 1992; Turner and O'Neill, 1992).

In this second part of the chapter, I address the specific problem of selecting a spatial scale for averaging complex systems. I illustrate with a predator–prey model that fundamental dynamical properties, such as dimensionality and degree of determinism, vary with scale. This variation can be characterized with methods at the interface of dynamical systems and time-series analysis, providing a basis to select a scale for aggregation. But scale selection is only the first step in model simplification. The problem then shifts to deriving an approximation for the dynamics of macroscopic quantities. Here, nonlinearity comes into play by enhancing heterogeneity and allowing fine-scale spatial structure to influence the dynamics at coarser scales.

Determinism, Dimensionality, and Scale in a Predator–Prey Model

To address the spatial scale for averaging, I chose a model that provides a useful metaphor for more realistic systems (Durrett and Levin, 2000). Thus, in its "idealized" form, the model already incorporates elements of nonlinearity, stochasticity, and local interactions. It is an individual-based predator–prey system in which the interactions among predators and prey occur in a defined spatial neighborhood. This antagonistic interaction can display a variety of local dynamics, including oscillations and extinctions. Over a landscape, persistence and spatiotemporal patterns reflect the intricate interplay of these local fluctuations. Thus, the model in its qualitative dynamics captures the essence of many other ecological systems, particularly those for host–parasite and host–parasitoid interactions (Comins et al., 1992). The spatial coupling of local fluctuations appears important to the dynamics of epidemics (Grenfell and Harwood, 1997); predators and their prey (Bascompte et al., 1997); and, more generally, of any system whose local feedbacks preclude local equilibria. In the marine environment, an interesting example can be found in benthic systems. In these systems, physical disturbance plays an important role by determining the renewal rate of space, often a limiting resource, thereby resetting local species succession. Although disturbance is generally viewed as an external forcing, there is evidence for biological feedbacks affecting its rate. For example, in the rocky intertidal, the barnacle hummocks that form when recruitment is high become unstable and susceptible to removal by wave action (Gaines and Roughgarden, 1985); and the crowding of mussels leads to the disruption of individuals and the formation of focal points for erosion and patch creation (Paine and Levin, 1981). Note that I have so far avoided a precise definition of "local" fluctuations. As illustrated below, the character of fluctuations in nonlinear systems is greatly modified by the spatial scale of sampling.

The predator–prey model follows the fate of individual predators and their prey in continuous time and two-dimensional space. Space consists of a lattice in which each site is either occupied by a predator, occupied by a prey, or empty (Figure 5). The state of a site changes in time according

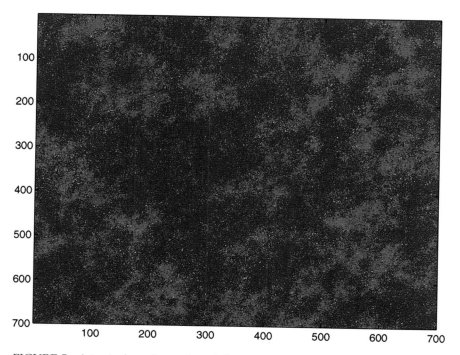

FIGURE 5 A typical configuration of the predator–prey model. The size of the lattice is 700×700. Empty sites are coded in blue, prey sites in red, and predator sites in yellow. Note that the prey forms clusters. These clusters continuously change as they form and disappear through the dynamic interplay of births and deaths. SOURCE: Pascual and Levin (1999b).

to the following processes: Predators hunt for prey by searching within a neighborhood of a prescribed size. Only predators that find prey can reproduce, and they do so at a specified rate. Predators that do not find prey are susceptible to starvation. Prey reproduce locally only if a neighboring site is empty. There is movement through mixing: All neighboring sites exchange state at a constant rate (for details see Durrett and Levin, 2000). In the model, nonlinearity results from predation and birth rates that depend on local densities. Stochasticity is demographic, representing the uncertainty in the fate of any single individual, and is implemented through rates that specify probabilities for the associated events to happen in a given interval of time.[5]

Simulations have shown that the spatiotemporal dynamics of the model perform an intricate dance as clusters or patches form and disappear (Pascual and Levin, 1999b) (Figure 5). As a result, the dynamics of population densities—our macroscopic quantity—change character with the sampling scale (Figure 6): At small size, stochasticity prevails (panel A); at an intermediate size, the dynamics appear less jagged, more continuous, or smooth (panel B); at a sufficiently large size, small fluctuations around a steady state result from averaging local dynamics that are out of phase (panel C).

A more precise description of these changes is obtained by quantifying how the degree of determinism varies with scale. By definition, the equations of a deterministic model express a causal and therefore predictable relationship between the past and the future: Given exact initial conditions, they completely specify the future state of the system. (Of course, in practice, measurement and dynamical noise are seldom if never absent, limiting predictability in chaotic systems.) When explicit equations are unavailable (as is the case here for densities), the degree of determinism can be evaluated from data through the (short-term) prediction accuracy of a prediction algorithm. The basic steps behind a prediction algorithm are shown in Figure 7. The starting point is by now familiar: Phase-space trajectories are "reconstructed" from a time series of population densities. The local geometry of these trajectories is then used to produce a short-term forecast. In a sense, the algorithm and the associated data constitute an "implicit model" whose (short-term) prediction accuracy reveals the prevalence of determinism versus noise in the dynamics (Kaplan and Glass, 1995).

Prediction accuracy (or predictability) is evaluated by comparing the mean prediction error to the variance of the time series (Kaplan and Glass, 1995) with a quantity such as

$$\text{Prediction} - r^2 = \frac{1 - \text{Mean prediction error}}{\text{Variance of data}} \qquad (4)$$

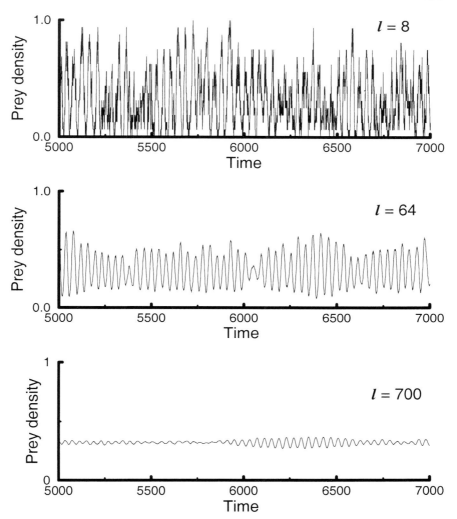

FIGURE 6   Prey densities in windows of different size. Prey densities are computed in a square window of area equal to l^2. The panels show the temporal behavior of prey density in windows of increasing area: (A), $l^2 = 8 \times 8$; (B), $l^2 = 64 \times 64$; (C), $l^2 = 700 \times 700$. SOURCE: Pascual and Levin (1999b).

If Prediction $- r^2$ is close to 0, then the mean prediction error is large with respect to the variance. In this case, the prediction algorithm provides a poor model for the data, and noise is prevalent in the dynamics. If Prediction $- r^2$ is close to 1, then the mean prediction error is small; the predic-

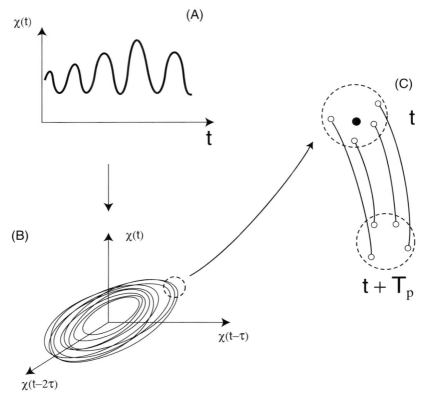

FIGURE 7   The basic steps of a prediction algorithm. From a time series for a
single variable, $x(t)$ (in A), phase-space dynamics are reconstructed using time-
delay coordinates (in B). For each point on the reconstructed trajectory, a prediction is then obtained as follows: (1) the algorithm chooses a given number k of
near neighbors [the open dots in (C) at time t]; (2) the algorithm follows the trajectory of each of these neighbors for the prediction horizon T_p; (3) a forecast is computed from the coordinates of these new points in ways that vary with the specific
algorithm (for examples and details see Farmer and Sidorowich, 1987; Kaplan and
Glass, 1995; Sauer, 1993; Sugihara and May, 1990). A mean prediction error is
computed after repeating these steps for a large number of points.

tion algorithm accounts for a large fraction of the variance, and the dynamics are predominantly deterministic.

   Figure 8 shows that determinism increases with spatial scale (Pascual
and Levin, 1999b). More interestingly, most of the change in predictability occurs at small scales, and there is little to gain in terms of determinism by considering windows larger than $l \geq 64$. At this intermediate scale,

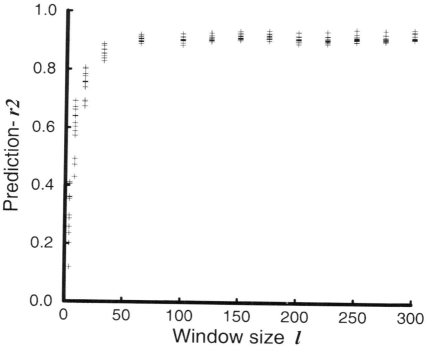

FIGURE 8 Predictability as a function of window size. Not surprisingly, for small window size Prediction $-r^2$ is close to 0, indicating the noisy character of the dynamics. At large window size, Prediction $-r^2$ is close to 1 because the dynamics of prey numbers are almost constant. Predictability has become high but trivial. More interestingly, the figure reveals the intermediate spatial scale at which high predictability is first achieved. (The basic shape of the curve remains unchanged for different parameters of the prediction algorithm; embedding dimension $d = 5$, 7, 9, 11; number of neighbors $k = d - 2, d, d+2$; lag $\tau = 11$; and prediction horizon $T_p = \tau$).

high predictability is achieved with a low number of variables (or number of lags). From the small individual scale to the intermediate scale l_c the dynamics lose their high-dimensional noisy character, with a significant reduction in the number of variables required to specify the state of the system as it changes in time. This reduction is confirmed by estimating the correlation dimension of the attractor (a measure of its dimensionality) (Pascual and Levin, 1999b).[6]

The intermediate scale $l_c = 64$ provides a natural size at which to model and sample population densities (Keeling et al., 1997; Rand and Wilson, 1995), not only because the dynamics have an important deter-

ministic component, but also because the model would provide information on the predator–prey interaction and its oscillatory nature. Deterministic equations for densities would have to capture the irregular fluctuations of predator and prey. One further, and somewhat unexpected,
property of the dynamics at scale l_c underscores that this is not an easy
matter. This property is sensitive to initial conditions, the trademark of
chaos: Arbitrarily small differences in initial conditions are amplified as
trajectories diverge exponentially in phase space (Figure 9).

The significance of sensitivity to initial conditions is not related to
chaos per se, but to the problem of deriving equations for population densities. The simpler candidate for such equations would be a "mean-field"
model derived by assuming that space is well mixed and therefore not
important. Clearly, such a model, a two-dimensional predator–prey system of differential equations, cannot exhibit sensitivity to initial conditions. For the specific parameters used here, this temporal model exhibits
periodic behavior, providing a poor approximation for the dynamics of
densities not only at scale l_c but also at larger scales. Thus, a model that
ignores spatial detail entirely is inadequate: The dynamics of our macroscopic quantity are affected by the fine-scale spatial clustering.

To better approximate the dynamics of densities at scale l_c, two extensions of simple predator–prey equations come to mind (Pascual and Levin,

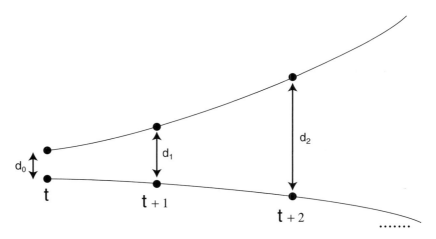

FIGURE 9   Sensitivity to initial conditions. To probe for sensitivity to initial
conditions at scale l_c, a time series of prey densities is used to reconstruct the
attractor, and the fate of nearby points is examined as follows. For each point, its
closest neighbor is found. The distances among the two respective trajectories are
recorded as time passes. Pascual and Levin (1999b) show that, on average, the
trajectories of nearby points diverge exponentially for small times.

1999b). The first one, of course, consists of incorporating the spatial dimension. Space is likely to be important because l_c falls close to the typical size of the clusters, a scale at which the dynamics of adjacent windows are likely to influence each other in non-negligible ways. There are indeed examples of spatial predator–prey models with chaotic dynamics, such as reaction-diffusion equations in which random movement couples local oscillations (Pascual, 1993; Sherratt et al., 1997). But chaos in these systems requires either specific initial conditions, representing the invasion of prey by predators (Sherratt et al., 1997), or the presence of an environmental gradient (Pascual, 1993). A second but more speculative possibility is that, beyond spatial structure, noise itself—the variability at scales smaller than l_c plays an important role in the dynamics, even at a scale where determinism is prevalent. In this case, deterministic equations that do not incorporate such variability would provide neither a template nor an adequate simplification for the dynamics of densities in window size l_c. These alternatives remain to be explored.

Some Open Areas

Investigations of scaling in spatial stochastic models for antagonistic interactions, such as the predator–prey system presented here, have only recently been initiated (deRoos et al., 1991; Keeling et al., 1997; Pascual and Levin, 1999b; Rand and Wilson, 1995). Along with results for specific models, these studies will develop general insights and approaches for simplification and scaling in systems with local cycles—systems whose dynamics in a range of spatial scales differ from random fluctuations around an equilibrium. These general results will find application in a variety of ecological systems. Likely candidates are systems that include host–parasite, host–parasitoid, or predator–prey interactions, as well as those where physical disturbance depends on local biological conditions. In the marine environment, predation and physical disturbance play an important role both in the plankton and in the benthos.

In ecology, the application of approaches at the interface of dynamical systems and time-series analysis to problems of aggregation has also only recently begun (Keeling et al., 1997; Little et al., 1996; Pascual and Levin, 1999b). These approaches should be useful for selecting appropriate levels of spatial as well as functional aggregation. Beyond scale selection, however, nonlinear time-series models have also an unexplored potential for the development of empirical methods to scale dynamics.

In the derivation of approximations for dynamics at coarser scales, emphasis so far has been on mathematical approaches (see Levin and Pacala, 1997, for a review). Because such formal derivations may prove impractical if not impossible in many models tailored to specific systems, there is a need for more empirical methods (e.g., Deutschman, 1996). Here, models that are flexible enough to incorporate what we know, but leave unspecified what we do not know, may prove useful.[7] For example, we could specify variables and relationships among variables, but not the exact functional form of these relationships.

Finally, we may see the development of methods that take advantage of descriptive scaling laws. Power laws relating a quantity of interest to the scale on which it is measured often describe elaborate patterns in nature (Mandelbrot, 1983; for ecological examples, see Hastings and Sugihara, 1993; Pascual et al., 1995). Such static descriptions, developed in the well-known field of fractals, may also prove useful to the scaling of complex systems in a dynamical context.

IN CLOSING

I started by presenting nonlinearity and the associated interplay of variability across scales as major roadblocks to the understanding and prediction of dynamics in ecological systems. Somewhat paradoxically, however, nonlinearity itself provides a way around these limitations. As the predator–prey model and other examples in the literature illustrate, nonlinear systems exhibit dynamics with perplexing order. This order, also known as self-organization, is reflected in spatiotemporal patterns that can be described with many fewer variables than those of the original system (Kauffman, 1993). One way to achieve this reduction in complexity is to exploit relationships between scales of description and dynamical properties, including dimensionality and predictability.

At levels of description for which the dynamics of ecological interactions are well captured by low-dimensional systems, we can also hope to better understand and predict responses to environmental forcings, but not without a broader perspective that encompasses nonlinearity.

In the end, in a nonlinear world there are alternatives, both fascinating and open for exploration, to abandoning the "Map of the Empire" to the "Inclemencies of the Sun and of the Winters."

NOTES

1. This disturbing character of chaos is illustrated in the title *Example of a Mathematical Deduction Forever Unusable* (Dunhem, 1906) described in Ruelle (1991).

2. Implementations include feed-forward neural networks or local predictors (for examples see Ellner and Turchin, 1995, or Ellner et al., 1998).

3. E_t can denote a vector time series when multiple forcings or lagged values of a forcing are considered.

4. The original text: '*Se le cose stanno così, il modello dei modelli vagheggiato da Palomar dovrà servire a ottenere dei modelli trasparenti, diafani, sottili come ragnatele; magari addirittura a dissolvere i modelli . . .*' —Italo Calvino, "Palomar."

5. Specifically, an event occurs at times of a Poisson process with the specified rate.

6. The estimated correlation dimension is 5.6, and the analysis shows that $d = 7$ is sufficient to reconstruct the attractor (Pascual and Levin, 1999b).

7. An example would be approaches that bring together studies of scaling and inverse methods.

REFERENCES

Abrams, P. A., B. A. Menge, G. G. Mittelbach, D. A. Spiller, and P. Yodzis. 1996. The role of indirect effects in food webs. In G. A. Polis and K. O. Winemiller, eds., *Food Webs: Integration of Patterns and Dynamics.* New York: Chapman & Hall.

Anderson, R. M., and R. M. May. 1991. *Infectious Diseases of Humans: Dynamics and Control.* Oxford, U.K.: Oxford University Press.

Andrewartha, H. G., and L. C. Birch. 1954. *The Distribution and Abundance of Animals.* Chicago: University of Chicago Press.

Armstrong, R. A. 1994. Grazing limitation and nutrient limitation in marine ecosystems: Steady-state solutions of an ecosystem model with multiple food chains. *Limnology and Oceanography* 39(3):597–608.

Bascompte, J., R. V. Solé, and N. Martinez. 1997. Population cycles and spatial patterns in snowshoe hares: An individual-oriented simulation. *Journal of Theoretical Biology* 187: 213–222.

Birch, L. C. 1958. The role of weather in determining the distribution and abundance of animals. *Cold Spring Harbor Symposia on Quantitative Biology* 22:203–218.

Bolker, B. M., and B. T. Grenfell. 1996. Impact of vaccination on the spatial correlation and dynamics of measles epidemics. *Proceedings of the National Academy of Sciences of the United States of America* 93:12,648–12,653.

Bolker, B. M., and S. W. Pacala. 1997. Using moment equations to understand stochastically driven spatial pattern formation in ecological systems. *Theoretical Population Biology* 52:179–197.

Borges, J. L. 1964. *Dreamtigers.* Austin: University of Texas Press.

Botsford, L. W., J. C. Castilla, and C. H. Peterson. 1997. The management of fisheries and marine ecosystems. *Science* 277:509–514.

Bouma, M. J, and H. J. van der Kaay. 1994. Epidemic malaria in India and the El Niño Southern Oscillation. *Lancet* 344:1638–1639.

Bouma, M. J., H. E. Sondorp, and H. J. van der Kaay. 1994. Climate change and periodic epidemic malaria. *Lancet* 343:1440.

Caperon, J. 1969. Time lag in population growth response in *Isochrysis galbana* to a variable nitrate environment. *Ecology* 50(2):188–192.

Casdagli, M. 1992. A dynamical systems approach to modeling input-outputs systems. In M. Casdagli and S. Eubank, eds., *Nonlinear Modeling and Forecasting.* New York: Addison-Wesley.

Caswell, H., and M. A. John. 1992. From the individual to the population in demographic models. In D. L. DeAngelis and L. J. Gross, eds., *Individual-based Models and Approaches in Ecology. Populations, Communities and Ecosystems.* New York: Chapman & Hall.

Clements, F. E. 1936. Nature and structure of the climax. *Journal of Ecology* 24:252–284.

Cohen, J. E. 1996. Unexpected dominance of high frequencies in chaotic nonlinear population models. *Nature* 378:610–612.

Colwell, R. R. 1996. Global climate and infectious disease: The cholera paradigm. *Science* 274:2025–2031.

Colwell, R. R., R. J. Seidler, J. Kaper, S. W. Joseph, S. Garges, H. Lockman, D. Maneval, H. Bradford, N. Roberst, E. Remmers, I. Huq, and A. Huq. 1981. Occurrence of *Vibrio cholerae* serotype O1 in Maryland and Louisiana estuaries. *Applied and Environmental Microbiology* 41:555–558.

Comins, H. N., M. P. Hassell, and R. M. May. 1992. The spatial dynamics of host-parasitoid systems. *Journal of Animal Ecology* 61:735–748.

Cushing, D. H., ed. 1983. *Key papers on fish populations.* Oxford, U.K.: IRL Press.

Dawson, T. E., and F. S. I. Chapin. 1993. Grouping plants by their form-function characteristics as an avenue for simplification in scaling between leaves and landscapes. In J. R. Ehleringer and C. B. Field, eds., *Scaling Physiological Processes: Leaf to Globe.* San Diego, Calif.: Academic Press.

Denman, K. L., and T. M. Powell. 1984. Effects of physical processes on planktonic ecosystems in the coastal ocean. *Oceanography and Marine Biology Annual Review* 22:125–168.

deRoos, A. M., E. McCauley, and W. W. Wilson. 1991. Mobility versus density-limited predator-prey dynamics on different spatial scales. *Proceedings of the Royal Society of London Series B* 246(1316):117–122.

deRoos, A. M., O. Diekman, and J. A. J. Metz. 1992. Studying the dynamics of structured population models: A versatile technique and its application to *Daphnia. American Naturalist* 139(1):123–147.

Deutschman, D. H. 1996. *Scaling from trees to forests: The problem of relevant detail.* Ph.D. Thesis, Cornell University.

Droop, M. R. 1966. Vitamin B_{12} and marine ecology. III. An experiment with a chemostat. *Journal of the Marine Biology Association of the United Kingdom* 46:659–671.

Dunhem, P. 1906. *La Théorie Physique: Son Objet et sa Structure.* Paris: Chevalier et Rivière.

Durrett, R., and S. A. Levin. 2000. Lessons on pattern formation from planet WATOR. *Journal of Theoretical Biology.*

Dwyer, R. L., and K. T. Perez. 1983. An experimental examination of ecosystem linearization. *American Naturalist* 121(3):305–323.

Dwyer, R. L., S. W. Nixon, C. A. Oviatt, K. T. Perez, and T. J. Smayda. 1978. Frequency response of a marine ecosystem subjected to time-varying inputs. In J. H. Thorp and J. W. Gibbons, eds., *Energy and Environmental Stress in Aquatic Ecosystems. U.S. DOE Symposium Series No. 48* (NTIS No. CONF-771114). Washington, D.C.: U.S. Department of Energy.

Ellner, S. P., and P. Turchin. 1995. Chaos in a 'noisy' world: New methods and evidence from time series analysis. *American Naturalist* 145:343–375.

Ellner, S. P., B. A. Bailey, G. V. Bobashev, A. R. Gallant, B. T. Grenfell, and D. W. Nychka. 1998. Noise and nonlinearity in measles epidemics: Combining mechanistic and statistical approaches to population modeling. *American Naturalist* 151:425–440.

Engbert, R., and F. R. Drepper. 1994. Qualitative analysis of unpredictability: A case study from childhood epidemics. In J. Grasman and G. van Straten, eds., *Predictability and Nonlinear Modeling in Natural Sciences and Economics.* Dordrecht, The Netherlands: Kluwer Academic.

Epstein, P. R., T. E. Ford, and R. R. Colwell. 1993. Marine ecosystems. In Health and Climate Change. *Lancet* 342:1216–1219.

Farmer, J. D. 1982. Chaotic attractors of an infinite-dimensional dynamical system. *Physica D* 4:366–393.

Farmer, J. D., and J. J. Sidorowich. 1987. Predicting chaotic time series. *Physics Review Letters* 59:845–848.

Gaines, S., and J. Roughgarden. 1985. Larval settlement rate: A leading determinant of structure in an ecological community of the marine intertidal zone. *Proceedings of the National Academy of Sciences of the United States of America* 82:3707–3711.

Gleason, H. A. 1926. The individualistic concept of the plant association. *Bulletin of the Torrey Botanical Club* 53:7–26.

Goulden, C. E., and L. L. Hornig. 1980. Population oscillations and energy reserves in planktonic cladocera and their consequences to competition. *Proceedings of the National Academy of Sciences of the United States of America* 77:1716–1720.

Grenfell, B. T., and A. P. Dobson. 1995. *Ecology of Infectious Diseases in Natural Populations.* New York: Cambridge University Press.

Grenfell, B. T., and J. Harwood. 1997. (Meta)population dynamics of infectious diseases. *Trends in Ecology and Evolution* 12(10):395–399.

Hadamard, J. 1898. Les surfaces à curbures opposées et leurs lignes géodesiques. *Journal of Mathematical Analyses and Applications* 4:27–73 (reprinted in Oeuvres de Jacques Hadamard. 1968. Paris: Centre National de la Recherche Scientifique).

Hassell, M. P., G. H. Lawton, and R. M. May. 1976. Patterns of dynamical behavior in single-species populations. *Journal of Animal Ecology* 45:471–486.

Hastings, A., C. L. Hom, S. P. Ellner, P. Turchin, and H. C. J. Godfray. 1993. Chaos in ecology: Is mother nature a strange attractor. *Annual Review of Ecology and Systematics* 24:1–33.

Hastings, H. M., and G. Sugihara. 1993. *Fractals: A User's Guide for the Natural Sciences.* New York: Oxford University Press.

Haury, L. R., J. A. McGowan, and P. H. Wiebe. 1978. Patterns and processes in the time-space scales of plankton distribution. In J. H. Steele, ed., *Spatial Pattern in Plankton Communities.* New York: Plenum.

Hay, M. 1994. Species as noise in community ecology: Do seaweeds block our view of the kelp forest? *Trends in Ecology and Evolution* 9(11):414–416.

Huq, A., E. B. Small, P. A. West, M. I. Huq, R. Rahman, and R. R. Colwell. 1983. Ecological relationships between *Vibrio cholerae* and planktonic crustacean copepods. *Applied and Environmental Microbiology* 45(1):275–283.

Huston, M., D. L. DeAngelis, and W. Post. 1988. New computer models unify ecological theory. *Biosciences* 38:682–691.

Inoue, M., and H. Kamifukumoto. 1984. Scenarios leading to chaos in a forced Lotka-Volterra model. *Progress in Theoretical Physics* 71(5):931–937.

Judson, O. P. 1994. The rise of the individual-based model in ecology. *Trends in Ecology and Evolution* 9(1):9–14.

Kaplan, D., and L. Glass. 1995. *Understanding Nonlinear Dynamics.* New York: Springer-Verlag.

Kauffman, S. A. 1993. *The Origins of Order. Self-organization and Selection in Evolution.* New York: Oxford University Press.

Keeling, M. J., I. Mezic, R. J. Hendry, J. McGlade, and D. A. Rand. 1997. Characteristic length scales of spatial models in ecology via fluctuation analysis. *Philosophical Transactions of the Royal Society of London Series B* 352:1589–1601.

King, A. W. 1992. Translating models across scales in the landscape. In C. G. Jones and J. H. Lawton, eds., *Linking Species and Ecosystems.* New York: Chapman & Hall.

Kingsland, S. E. 1985. *Modeling Nature. Episodes in the History of Population Ecology.* Chicago: University of Chicago Press.

Kot, M., W. M. Schaffer, G. L. Truty, D. J. Graser, and L. F. Olsen. 1988. Changing criteria for imposing order. *Ecological Modelling* 43:75–110.

Kot, M., G. S. Sayler, and T. W. Schultz. 1992. Complex dynamics in a model microbial system. *Bulletin of Mathematical Biology* 54(4):619–648.

Levin, S. A. 1992. The problem of pattern and scale in ecology. *Ecology* 73(6):1943–1967.

Levin, S. A., and R. Durrett. 1996. From individuals to epidemics. *Philosophical Transactions of the Royal Society of London Series B* 351:1615–1621.

Levin, S. A., and S. W. Pacala. 1997. Theories of simplification and scaling in ecological systems. In D. Tilman and P. Kareiva, eds., *Spatial Ecology: The Role of Space in Population Dynamics and Interspecific Interactions.* Princeton, N.J: Princeton University Press.

Levin, S. A., and M. Peale. 1996. Beyond extinction: Rethinking biodiversity. Bulletin of the Santa Fe Institute (Winter 1995–1996):2–4.

Levin, S., and L. A. Segel. 1976. Hypothesis for origin of planktonic patchiness. *Nature* 259:659.

Levin, S., and L. A. Segel. 1985. Pattern generation in space and aspect. *SIAM Review* 27(1):45–67.

Little, S. A., S. P. Ellner, M. Pascual, M. Neubert, D. T. Kaplan, T. Sauer, A. Solow, and H. Caswell. 1996. Detecting nonlinear dynamics in spatio-temporal systems: Examples from ecological models. *Physica* D 96:321–333.

Lorenz, E. N. 1963. Deterministic nonperiodic flow. *Journal of Atmospheric Science* 20:130–141.

Mandelbrot, B. B. 1983. *The Fractal Geometry of Nature.* New York: Freeman.

May, R. 1974. Biological populations with non-overlapping generations: Stable points, stable cycles and chaos. *Science* 186:645–647.

McCauley, E., and W. W. Murdoch. 1987. Cyclic and stable populations: Plankton as a paradigm. *American Naturalist* 129:97–121.

Mollison, D., ed. 1995. *Epidemic Models: Their Structure and Relation to Data.* New York: Cambridge University Press.

Murray, J. D. 1989. *Mathematical Biology.* New York: Springer-Verlag.

Nicholson, A. J. 1958. Dynamics of insect populations. *Annual Review of Entomology* 3:107–136.

Nisbet, R. M., and W. S. C. Gurney. 1985. Fluctuation periodicity, generation separation, and the expression of larval competition. *Theoretical Population Biology* 28:150–180.

Okubo, A. 1980. Diffusion and Ecological Problems: Mathematical Problems. *Biomathematics, Vol. 10.* New York: Springer-Verlag.

Olsen, L. F., G. L. Truty, and W. M. Schaffer. 1988. Oscillations and chaos in epidemics: A nonlinear dynamic study of six childhood diseases in Copenhagen, Denmark. *Theoretical Population Biology* 33:344–370.

Pacala, S. W., and D. Deutschman. 1996. Details that matter: The spatial distribution of individual trees maintains forest ecosystem function. *Oikos* 74:357–365.

Paine, R. T., and S. A. Levin. 1981. Intertidal landscapes: Disturbance and the dynamics of pattern. *Ecological Monographs* 51(2):145–178.

Pascual, M. 1993. Diffusion-induced chaos in a spatial predator-prey system. *Proceedings of the Royal Society of London Series B* 251:1–7.

Pascual, M. 1994. Periodic response to periodic forcing of the Droop equations for phytoplankton growth. *Journal of Mathematical Biology* 32:743–759.

Pascual, M., and H. Caswell. 1997a. From the cell-cycle to population cycles in phytoplankton-nutrient interactions. *Ecology* 78(3):897–912.

Pascual, M., and H. Caswell. 1997b. Environmental heterogeneity and biological pattern in a chaotic predator-prey system. *Journal of Theoretical Biology* 185:1–13.

Pascual, M., and S. A. Levin. 1999a. Spatial scaling in a benthic population model with density-dependent disturbance. *Theoretical Population Biology* 56:106–122.

Pascual M., and S. A. Levin. 1999b. From individuals to population densities: Searching for the intermediate scale of nontrivial determinism. *Ecology* 80(7):2225–2236.

Pascual, M., F. A. Ascioti, and H. Caswell. 1995. Intermittency in the plankton: A multifractal analysis of zooplankton biomass variability. *Journal of Plankton Research* 17(6):1209–1232.

Patz, J. A., P. R. Epstein, T. A. Burke, and J. M. Balbus. 1996. Global climate change and emerging infectious diseases. *Journal of the American Medical Association* 275(3):217–223.

Pickett, J. M. 1975. Growth of *Chlorella* in a nitrate-limited chemostat. *Plant Physiology* 55:223–225.

Pickett, S. T. A., and P. S. White. 1985. *The Ecology of Natural Disturbance and Patch Dynamics.* Orlando, Fla: Academic Press.

Poincaré, H. 1908. *Science et Méthode.* Paris: Ernest Flammarion.

Rand, D. A., and H. B. Wilson. 1991. Chaotic stochasticity: A ubiquitous source of unpredictability in epidemics. *Proceedings of the Royal Society of London Series B* 246:179–184.

Rand, D. A., and H. B. Wilson. 1995. Using spatio-temporal chaos and intermediate-scale determinism to quantify spatially extended ecosystems. *Proceedings of the Royal Society of London Series B* 259:111–117.

Rastetter, E. B., and G. R. Shaver. 1995. Functional redundance and process aggregation: Linking ecosystems and species. In C. G. Jones and J. H. Lawton, eds., *Linking Species and Ecosystems.* New York: Chapman & Hall.

Rastetter, E. B., A. W. King, B. J. Cosby, G. M. Hornberger, R. V. O'Neill, and J. E. Hobbie. 1992. Aggregating fine-scale ecological knowledge to model coarser-scale attributes of ecosystems. *Ecological Applications* 2(1):55–70.

Rinaldi, S., S. Muratori, and Y. Kuznetzov. 1993. Multiple attractors, catastrophes, and chaos in seasonally perturbed predator-prey communities. *Bulletin of Mathematical Biology* 55:15–36.

Rothschild, B. J. 1986. *Dynamics of Marine Fish Populations.* Cambridge, Mass.: Harvard University Press.

Ruelle, D., 1991. *Chance and Chaos.* Princeton, N.J.: Princeton University Press.

Ruelle, D., and F. Takens. 1971. On the nature of turbulence. *Communications in Mathematical Physics* 20:167–192; 23:343–344.

Sauer, T. 1993. Time series prediction by using delay coordinate embedding. In A. S. Weigend and N. A. Gershenfeld, eds., *Time Series Prediction: Forecasting the Future and Understanding the Past. SFI Studies in the Sciences of Complexity, Proceedings Vol. XV.* Reading, Mass.: Addison Wesley.

Schaffer, W. M. 1988. Perceiving order in the chaos of nature. In M. S. Boyce, ed., *Evolution of Life Histories in Mammals, Theory and Pattern.* New Haven, Conn.: Yale University Press.

Schaffer, W. M., and M. Kot. 1985. Do strange attractors govern ecological systems? *Bioscience* 35:342–350.

Schaffer, W. M., and M. Kot. 1986. Chaos in ecological systems: The coals that Newcastle forgot. *Trends in Ecology and Evolution* 1:58–63.

Schaffer, W. M., L. F. Olsen, G. L. Truty, and S. L. Fulmer. 1990. The case for chaos in childhood epidemics. In S. Krasner, ed., *The Ubiquity of Chaos.* Washington, D.C.: American Association for the Advancement of Science.

Sherratt, J. A., B. T. Eagan, and M. A. Lewis. 1997. Oscillations and chaos behind predator-prey invasion: Mathematical artifact or ecological reality? *Philosophical Transactions of the Royal Society of London Series B* 352:21–38.

Sidorowich, J. J. 1992. Repellors attract attention. *Nature* 355:584–585.

Smith, F. E. 1961. Density dependence in the Australian thrips. *Ecology* 42:403–407.

Solbrig, O. T. 1993. Plant traits and adaptive strategies: Their role in ecosystem function. In E. D. Schulze and H. A. Mooney, eds., *Biodiversity and Ecosystem Function.* Berlin: Springer-Verlag.

Star, J. L., and J. J. Cullen. 1981. Spectral analysis: A caveat. *Deep-Sea Research* 28A:93–97.

Steele, J. H. 1978. *Spatial Pattern in Plankton Communities.* New York: Plenum.

Steele, J. H. 1988. Scale selection for biodynamic theories. In B. J. Rothschild, ed., *Toward a Theory on Biological-Physical Interactions in the World Ocean.* Dordrecht, The Netherlands: Kluwer Academic.

Steele, J. H., S. R. Carpenter, J. E. Cohen, P. K. Dayton, and R. E. Ricklefs. 1993. Comparing terrestrial and marine ecological systems. In S. A. Levin, T. M. Powell, and J. H. Steele, eds., *Patch Dynamics. Lecture Notes in Biomathematics 96.* New York: Springer-Verlag.

Steneck, R. S., and M. N. Dethier. 1994. A functional group approach to the structure of algal-dominated communities. *Oikos* 69:476–498.

Sugihara, G. 1994. Nonlinear forecasting for the classification of natural time series. *Philosophical Transactions of the Royal Society of London Series A* 348:477–495.

Sugihara, G. 1996. Out of the blue. *Nature* 378:559–560.

Sugihara, G., and R. M. May. 1990. Nonlinear forecasting as a way of distinguishing chaos from measurement error in time series. *Nature* 344:734–741.

Takens, F. 1981. Detecting strange attractors in turbulence. In D. Rand and L. S. Young, eds., *Dynamical Systems and Turbulence, Warwick 1980. Lecture Notes in Mathematics Vol. 898.* New York: Springer-Verlag.

Taubes, G. 1997. Apocalypse not. *Science* 278:1004–1006.

Tromp, S. W. 1963. *Medical Biometeorology. Weather, Climate and Living Organisms.* Amsterdam: Elsevier.

Tuljapurkar, S., and H. Caswell. 1996. *Structured Population Models.* New York: Chapman & Hall.

Turchin, P., and A. D. Taylor. 1992. Complex dynamics in ecological time series. *Ecology* 73:289–305.

Turner, M. G., and R. V. O'Neill. 1992. Exploring aggregation in space and time. In C. G. Jones and J. H. Lawton, eds., *Linking Species and Ecosystems.* New York: Chapman & Hall.

Williams, F. M. 1971. Dynamics of microbial populations. In B. C. Patten, ed., *Systems Analysis and Simulation in Ecology, Vol. 1.* New York: Academic Press.

World Health Organization (WHO). 1990. Potential Health Effects of Climatic Change. WHO/PEP/90.10. Geneva: WHO.

13

Currents of Change: The Ocean's Role in Climate

Stefan Rahmstorf

In 1751, the captain of an English slave-trading ship made a historic discovery. While sailing at latitude 25°N in the subtropical North Atlantic Ocean, Captain Henry Ellis lowered a "bucket sea-gauge," devised and provided for him by a British clergyman, the Reverend Stephen Hales, down through the warm surface waters into the deep. By means of a long rope and a system of valves, water from various depths could be brought up to the deck, where its temperature was read from a built-in thermometer. To his surprise Captain Ellis found that the deep water was icy cold.

He reported his findings to Reverend Hales in a letter: "The cold increased regularly, in proportion to the depths, till it descended to 3900 feet: from whence the mercury in the thermometer came up at 53 degrees (Fahrenheit); and tho' I afterwards sunk it to the depth of 5346 feet, that is a mile and 66 feet, it came up no lower."

These were the first ever recorded temperature measurements of the deep ocean. And they revealed what is now known to be a fundamental and striking physical feature of all the oceans: Deep water is always cold (Warren and Wunsch, 1981). The warm waters of the tropics and subtropics are confined to a thin layer at the surface; the heat of the Sun does not slowly warm up the depths as might be expected.

Ellis' letter to Hales suggests that he had no inkling of the far-reaching significance of his discovery. He wrote: "This experiment, which seem'd at first but mere food for curiosity, became in the interim very

Climate Research Department, Potsdam Institute for Climate Impact Research, Potsdam, Germany

useful to us. By its means we supplied our cold bath, and cooled our wines or water at pleasure; which is vastly agreeable to us in this burning climate" (Ellis, 1751).

In fact, Ellis had struck upon the first indication of the "thermohaline circulation," the system of deep ocean currents that circulates cold waters of polar origin around the planet, often referred to as the "Great Ocean Conveyor Belt."

But it was not until several decades later, in 1798, that another Englishman, Count Rumford, published a correct explanation for Ellis' "useful" discovery: "It appears to be extremely difficult, if not quite impossible, to account for this degree of cold at the bottom of the sea in the torrid zone, on any other supposition than that of *cold currents from the poles*; and the utility of these currents in tempering the excessive heats of these climates is too evident to require any illustration" (Thompson, 1798).

Now, 200 years later, using the most advanced supercomputers our century can provide, we are beginning to understand the intricate dynamics ruling the complex system of deep ocean circulation and, what Rumford found so evident, the role it plays in climate. It is a subject that may be of fundamental importance to our future.

AN OCEAN IN THE COMPUTER

My work is climate modeling; I simulate the currents of the world's oceans in a computer and investigate their transport of heat across the globe. The model I most frequently work with was developed at the Geophysical Fluid Dynamics Laboratory at Princeton and is used by many oceanographers around the world; I have adjusted it to best suit my experiments. The surface of the planet is divided into grid cells. My present model has 194 cells in longitude, 96 in latitude, and 24 vertical levels: altogether almost half a million grid points. At each point where there is ocean, the temperature and salinity of the water and the velocity of the currents are computed using basic hydrodynamic and thermodynamic equations for each time step that has been programmed. If the model is run for, say, 100 simulated years, roughly 100,000 time steps would be required. If this is multiplied by the number of grid points, it becomes clear why even the fastest supercomputers available will take quite some time to perform the huge calculations. In fact, to reach a steady state (or equilibrium) in the ocean circulation takes several thousand simulated years; so many calculations are necessary that it takes a supercomputer several weeks to perform them.

So what happens when such a simulation is run? First one has to specify a way to calculate the exchange of heat, freshwater (through evaporation, precipitation, and river runoff), and momentum (from the wind) at the ocean surface. This is called "forcing"—these *forces* drive the ocean circulation. The model then computes what kind of currents develop in the virtual ocean. The results of such a simulation, with a forcing based on the present-day climate, are illustrated in Figure 1. The computer model has reproduced all the major ocean currents known from shipboard measurements. The Gulf Stream and its extension toward Britain and Scandinavia (illustrated in the top panel) and the southward flow of North Atlantic Deep Water (NADW) out of the Atlantic (in the bottom

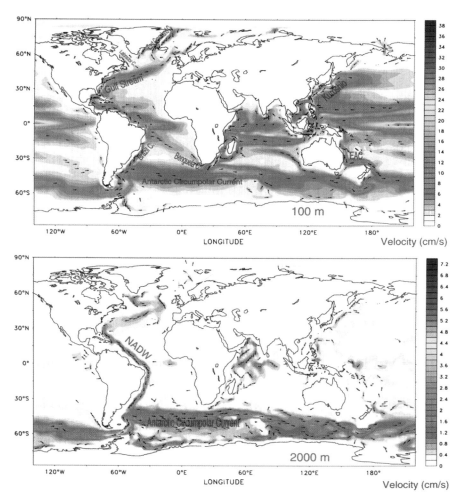

FIGURE 1 Current velocities in centimeters per second (color scale) in the ocean model. Top panel is at 100 m depth; bottom panel is 2,000 m depth.

panel) are of particular importance. These currents work together as a kind of conveyor belt bringing warmth to Europe.

THE CONVEYOR BELT: A CONTROVERSIAL CONCEPT

The concept of an ocean "conveyor belt" was first formed by Broecker (1987, 1991) of Columbia University to illustrate the idea that all the oceans in the world were connected through one coherent circulation system, which transported heat and salt[1] between them. Figure 2 shows a version of Broecker's famous sketch of 1987, depicting the system as a conveyor belt transporting warm water along the surface and cold water back through the depths.

Broecker believed that the global conveyor belt was driven by the atmosphere's transport of water vapor from the Atlantic basin to the Pacific—water evaporating from the Atlantic and raining down in the Pacific catchment area (see "Driving the Conveyor" below). His theory was that the strength of the conveyor belt flow was proportional to this vapor

FIGURE 2 The global conveyor belt after Broecker. SOURCE: Adapted from Broecker (1995).

transport. If the vapor transport were reduced (by less evaporation, for example), the entire system would slow down, just as a real conveyor belt does when its power is reduced.

It was a brilliant and provocative idea and "the global ocean conveyor belt" became a standard term for describing world ocean circulation in popular and scientific publications.

Some oceanographers, however, questioned the metaphor and the theory behind it, and in subsequent years a number of research papers have been published that give a more accurate picture of the way the ocean circulation functions. In the work I have been doing since 1991, I found that the circulation loops in the Atlantic and Pacific are only weakly connected and that the oceans do not respond as one system. *Within* the Atlantic Ocean, however, the circulation loop functions very much like a conveyor belt—warm water is transported north by a system of currents including the Gulf Stream to near Greenland, where it drops down to become the cold NADW flowing south (Figure 3).

And if extra freshwater is added in the computer model to the region near Greenland (as could happen with extra rainfall or the melting of ice in the real world), the whole system slows down as one, from the North Atlantic Drift down to the Benguela Current of the South Atlantic.

THE ROLE OF THE ATLANTIC CONVEYOR IN CLIMATE

The cold water discovered in the subtropical Atlantic by Ellis in 1751 was, as Rumford theorized, brought there by a current that had originated in the polar region. Temperature measurements in the real ocean and computer models show that there is a southward outflow of cold deep water from the Arctic throughout the Atlantic. This cold water is replaced by warm surface waters, which gradually give off their heat to the atmosphere as they flow northward toward Europe. This acts as a massive "central heating system" for all the land downwind.

The heat released by this system is enormous: It measures around 10^{15} W, equivalent to the output of a million large power stations. If we compare places in Europe with locations at similar latitudes on the North American continent, its effect becomes obvious. Bodö in Norway has average temperatures of $-2°C$ in January and $14°C$ in July; Nome, on the Pacific Coast of Alaska at the same latitude, has a much colder $-15°C$ in January and only $10°C$ in July (Weaver, 1995). And satellite images show how the warm current keeps much of the Greenland–Norwegian Sea free of ice even in winter, despite the rest of the Arctic Ocean, even much further south, being frozen.

If the Atlantic conveyor belt circulation is switched off in a computer model, a different climate forms in the virtual world. There is little change

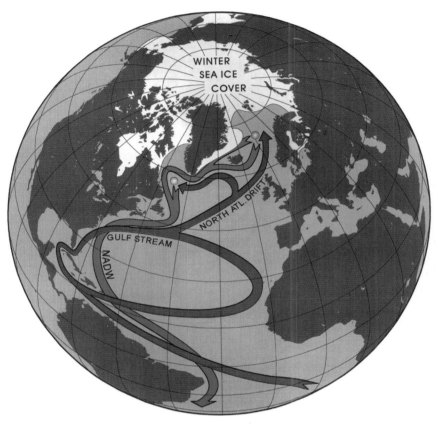

FIGURE 3 The Atlantic conveyor belt. Orange circles show the regions of convection in the Greenland–Norwegian and Labrador Seas. The outflow of NADW is shown in blue.

in ocean temperatures near the equator, but the North Atlantic region becomes much colder than it is in reality, and the South Atlantic and other parts of the Southern Hemisphere become warmer. This experiment reveals that the Atlantic circulation moves heat from the South Atlantic below the equator across the tropics to the North Atlantic—the heat is not coming directly out of the tropical region.

So Rumford did not get it quite right: The ocean currents do not seem to do much to cool the "excessive heats" of the tropics, although they certainly play an important role in preventing excessive cold in Britain, Scandinavia, and the rest of Northern Europe.

Some of my colleagues have compared climates with and without the Atlantic conveyor belt in computer simulations using coupled ocean–atmosphere models (Manabe and Stouffer, 1988). These models show that, without the ocean heating, air temperatures would cool by up to 10°C averaged over a year. The chill is greatest near Scandinavia, but extends to a lesser extent across Europe and much of the Northern Hemisphere.

DRIVING THE CONVEYOR

What drives this remarkable circulation? Why does it occur only in the Atlantic, and why is it that the Pacific and Indian Oceans do not have similar heating systems?

In general, ocean currents are driven either by winds or by density differences. Density in the ocean depends on temperature and salinity, and the Atlantic conveyor is a *thermohaline* circulation (from the Greek words for "heat" and "salt"). When surface waters become dense enough, through cooling or becoming saltier (or a combination of these two factors), a mixing process takes place in which they sink and form deep water.

Count Rumford understood the basics of this mechanism and reported them in his 1798 publication. The deep mixing takes place because of convection; seawater from near the surface sinks down 2 km or more and then spreads at this depth. This is called "deep-water formation."

Because this convection takes place only at specific sites, the image of a plug hole in a bathtub is suggestive. At certain locations on the broad expanse of ocean, the surface waters descend into the deep. They do not exactly gurgle down the plug hole, rather they subside so slowly that it is hard to measure directly. In the present climate, deep water forms in the Greenland–Norwegian Sea and in the Labrador Sea (marked by orange circles in Figure 3). There are no deep convection sites in the North Pacific. The only other place where the global ocean's deep water is formed is near the Antarctic continent in the Southern Hemisphere.

At these convection sites, the water has become dense enough to push away the underlying deep water and sink because it has cooled—cold water is denser than warm. In the Indian Ocean, the water is too warm to sink; the ocean's basin does not extend very far north of the equator.

But cold is not the only factor. The waters of the North Pacific are cold, but even as they approach the freezing point they still lack the density that would enable them to sink down and thus drive a conveyor flow. This is because the North Pacific is less salty than the North Atlantic (around 32 compared with 35 parts per thousand in the northern North Atlantic), and salt is the second crucial factor in the density of ocean water.

In fact, the high salinity of the North Atlantic is the key to understanding how the thermohaline circulation works. For decades, two theories were widely accepted as explanations of what drives the flow of NADW. Both were frequently cited. At first glance they seem complimentary, but my work led me to examine them more closely.

Broecker reasoned that the global conveyor (see Figure 2) was driven by evaporation. He explained the salinity difference between the Atlantic and the Pacific as a result of excess freshwater evaporating from the Atlantic basin and being blown across into the Pacific catchment by easterly winds—what is known as vapor transport. In other words, water was evaporating from the Atlantic leaving its salt behind and raining back down as freshwater which diluted the Pacific.

So in Broecker's theory, the relatively high salinity of the North Atlantic water, coupled with the cold temperatures, created the convection "pumps" that drove the global conveyor. The salinity budget of the oceans was balanced by the conveyor transporting fresher Pacific water back to the North Atlantic.

If, the theory continues, the conveyor were to grind to a halt for some reason, then the salinity of the North Atlantic would start to rise, as evaporation would continue to leave salty water behind, although the inflow of fresher water would have stopped.

Systematic computer simulations by myself and other oceanographers found flaws in this theory. We discovered that the present thermohaline circulation could be maintained without any airborne transport of water vapor from the Atlantic to the Pacific, and even with a weak *reverse* vapor transport. And, surprisingly, within such a scenario the North Atlantic continued to be saltier than the North Pacific.

Finding the explanation for this paradox was simple: The conveyor does not need evaporation in order to operate; it transports salty water into the North Atlantic by itself, thus maintaining the high salinity there. It is a classic chicken and egg situation—the Atlantic conveyor functions because salinity is high in the North Atlantic, and salinity is high in the North Atlantic because the conveyor is functioning. It is a positive feedback that makes the conveyor a self-sustaining system.

This peculiar property had already been examined much earlier, in 1961, by the famous American oceanographer Henry Stommel in a simple but powerful conceptual model. One consequence of the positive feedback was that if it was somehow interrupted, the conveyor belt would grind to a permanent halt. In the absence of ongoing circulation, the North Atlantic's salinity would drop so much that no more deep water could form.

After I had been working on the thermohaline circulation for some time, I realized that there was a contradiction between these two theories—Broecker's evaporation-driven conveyor and Stommel's self-maintaining conveyor—which meant that both could not be true. Each was widely cited in oceanographic literature, both were accepted as valid, and yet on closer examination I found that they were mutually exclusive.

In Broecker's theory, the direction of the freshwater transport by the Atlantic conveyor was northward; in Stommel's theory it was southward. This is why, if the circulation halted, the salinity of the North Atlantic would *increase* according to Broecker, but *decrease* according to Stommel. Obviously this result was directly dependent on whether the upper, northward flowing branch of the Atlantic conveyor (see Figure 3) had a higher or lower salinity than the outflow of NADW.

Existing hydrographic measurements from the real ocean were unable to resolve this issue because of the complicated salinity layering of the Atlantic waters. Above the layer of NADW, northward flowing layers exist with both higher *and* lower salinities. From the data, it is not possible to identify which of the near-surface currents belong to the thermohaline conveyor and which are simply driven by winds. So I set the computer model to the task.

Through a series of model experiments, I found that in the model world the conveyor belt transport of freshwater is indeed southward and that salinity decreases throughout most of the Atlantic when the conveyor is shut down. There was net evaporation from the Atlantic in my model, as in the real world, but this had little effect on the functioning of the conveyor. Also, the freshwater loss to the atmosphere was not balanced by the conveyor, but rather by wind-driven, near-surface currents.

In this sense, Stommel's theory was the more accurate of the two. But his work, done in 1961, had been based on a very simple "box" model of a theoretical thermohaline circulation. It was limited to one hemisphere and driven by the density difference between the water of the tropics and that of the high latitudes. Model experiments and observational data demonstrate that the real Atlantic does not work like this, although some of my colleagues have tried to argue for a theory of two more or less disconnected cells, one in each hemisphere.

My model results showed the existence of one *cross-hemispheric* conveyor belt in the Atlantic, transporting heat from the Southern to the Northern Hemisphere, which is driven by the density differences between North and South Atlantic water. If the northern part of the Atlantic thermohaline circulation is slowed down by adding virtual freshwater to the model ocean, the whole system slows down, including the Benguela Current off South Africa in the Southern Hemisphere.

To overturn established beliefs is not always easy. In 1995, I wrote a paper pointing out the inherent contradiction in the two major thermoha-

line circulation theories and arguing for a southward direction of freshwater transport. I submitted it to a highly respected scientific journal. My
work was given the thumbs down: One of the two reviewers criticized
my results for not being new, saying everyone in the field had known all
this for a long time; the other wrote that my results were wrong and flying
in the face of all evidence. This outcome all too accurately emphasized
the problem—fortunately I was able to publish elsewhere (Rahmstrof,
1996).

UNSTABLE CURRENTS

There is another crucial difference between the two views of the conveyor. A circulation driven by evaporation *and* high-latitude cooling
would be very stable. But Stommel's self-maintaining conveyor depends
on precariously balanced forces: Cooling pulls in one direction, while the
input of freshwater from rain, snow, melting ice, and rivers pulls in the
other. This freshwater threatens to reduce the salinity and, therefore, the
density of the surface waters; only by a constant flushing away of the
freshwater and replenishing with salty water from the south does the conveyor survive. If the flow slows down too much, there comes a point
where it can no longer keep up and the conveyor breaks down.

A look at a simple stability diagram shows how this works (Figure 4).
The key feature is that there is a definite threshold (labeled *S* and called a
"bifurcation") for how much freshwater input the conveyor can cope with.
Such thresholds are typical for complex, nonlinear systems. This diagram
is based on Stommel's theory, adapted for the Atlantic conveyor, and on
global circulation model experiments (Rahmstorf, 1995).

Different models locate the present climate (*P*) at different positions
on the stability curve—for example, models with a rather strong conveyor
are located further left in the graph and require a larger increase in precipitation to push the conveyor "over the edge" (transition *a* in Figure 4).
The stability diagram is thus a unifying framework that allows us to understand different computer models and experiments.

My experiments also revealed another kind of threshold where the
conveyor can suddenly change. While the vulnerability in Stommel's
model arises from the large-scale transport of salt by the conveyor, this
new type of threshold depends on the vertical mixing in the convection
areas (e.g., Greenland Sea, Labrador Sea). If the mixing is interrupted,
then the conveyor may break down completely in a matter of years, or *the
locations of the convection sites may shift*. Such a shift in convection sites is
indicated as transition *b* in Figure 4.

Although the effect on climate of a cessation of Atlantic circulation
had come to be generally understood, until 1994 no one had considered

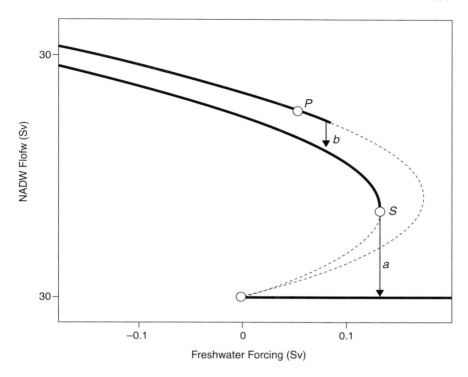

FIGURE 4   A schematic stability diagram shows how the flow of the conveyor depends on the amount of freshwater entering the North Atlantic. Units are 10^6 m^3/s on both axes.

the possibility of the shifting of convection sites (or "convective instabil-ity" as it is now known), let alone the consequences. That year, quite by chance, I discovered that such an occurrence would have a major effect: Regional climates would change radically in less than a decade.

The paper I published on the subject (Rahmstorf, 1994) struck a chord with paleoclimatologists, scientists who had been seeking explanations for the evidence they had found of abrupt climatic changes taking place thousands of years ago (see "The Past: A Roller Coaster Ride" below).

The speed at which these convective shifts take place is frightening. With Stommel's theory, once the freshwater threshold has been exceeded, the conveyor circulation slowly grinds to a halt over a century or more. But if a convective instability is triggered, within some few years the con-veyor stops transporting heat to the far north.

We do not yet know where these critical limits of convection are or what it would take to set off such an event. Current climate models are not powerful enough to resolve such regional processes clearly.

ICY TIMES

By looking back at past climates we can understand a little more about the effect of the North Atlantic circulation. The ancient icecaps of Greenland and Antarctica have preserved a unique and detailed record of the history of climate, layed down in year after year of snow that never melted, going back at least 100,000 years. Several cores have been drilled right through these mountain-high icecaps in recent years (Figure 5), and from the exact composition of the ice and enclosed air bubbles in different layers much information about the past climate can be recovered. Other valuable records of the past are contained in sea corals, tree rings, and in ancient pollen (pollen even reveals information about vegetation cover at different times in the planet's history). And cores taken from the sediments at the ocean bottom give a wealth of clues about past ocean circulation and climate.

From these data, it has been possible to reconstruct conditions at the

FIGURE 5   Scientists drilling an ice core in Greenland. Photo courtesy of Richard B. Alley, Pennsylvania State University.

height of the last Ice Age (the so-called "Last Glacial Maximum") around 21,000 years ago. Huge ice sheets, several kilometers thick, covered the northern parts of North America and Eurasia. In Europe, ice covered all of Scandinavia and reached down as far as Berlin. As far south as France, cold, dry steppe extended across the continent, stalked by mammoths.

Evidence points to the ocean's thermohaline circulation being quite different from today. Reconstructions show that the Atlantic conveyor did not nearly extend as far north. The convection regions were south of Iceland, and the water sank only to intermediate depths. The bottom of the North Atlantic was instead filled by waters of Antarctic origin which were pushing in from the south (Labeyrie et al., 1992).

In 1997, my colleagues and I at the Potsdam Institute for Climate Impact Research in Germany performed a computer simulation of the Ice Age climate (Ganopolski et al., 1998) (Figure 6), including both atmospheric and oceanic circulations—the first time such a simulation has been performed to our knowledge. Our model produced exactly the ocean circulation changes described above, and we could establish that the changes in the conveyor had a significant effect on Ice Age temperatures. We found that with the changes in the ocean circulation, the Northern Hemisphere was on average 9°C colder in the Ice Age than today, but if we experimented by deliberately preventing the ocean circulation changes, the temperatures were only 6°C colder.

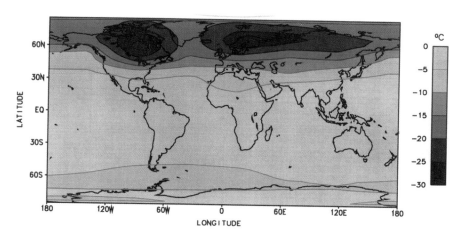

FIGURE 6 Cooling of surface temperatures at the height of the last Ice Age during Northern Hemisphere summer, as simulated by the Potsdam coupled climate model. SOURCE: Ganopolski et al. (1998).

It is a combination of such model simulations and further detective work on ice cores and similar data that will ultimately lead to a detailed understanding of the forces shaping the climate of our planet.

THE PAST: A ROLLER-COASTER RIDE

The coldness of the last Ice Age was punctuated by many sudden and erratic swings in the climate. The Greenland ice cores show sudden temperature shifts of about 5°C that happened over about a decade, but last for centuries (Figure 7). Many of these events were not local to Greenland, but had repercussions that have been detected as far afield as South America and New Zealand. When we look back over the history of climate, the past 10,000 years (the Holocene) appear as an unusually stable period. It is probably no coincidence that this is the time in which agriculture was invented and human civilization developed.

The cause of these rapid fluctuations puzzles paleoclimatologists. Subtle and gradual changes in the energy that the Earth receives from the Sun, due to wobbles in our orbit (the so-called Milankovich cycles), are the major reason for Ice Ages and other climate changes of the past. But why does the climate not respond in a smooth, gradual way? This is one of the greatest riddles in climatology.

Sediment cores from the sea bottom reveal that ocean currents were changing in sync with the weather over Greenland and other land areas. The shifts in ocean circulation (through the convective instability mechanism) together with instabilities in the large ice sheets may well be the culprits responsible for these erratic climate changes.

Even the relatively stable climate of the Holocene has not been an entirely smooth ride for humanity. An as yet unexplained cold snap oc-

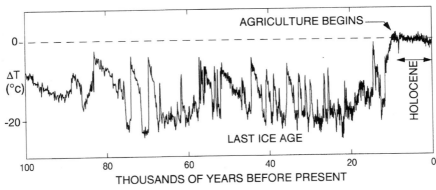

FIGURE 7   Temperature record of the past 100,000 years, derived from a Greenland ice core.

curred around 8,000 years ago—perhaps also caused by a change in ocean circulation. The warm "Holocene Optimum" followed—around 6,000 years ago—when the Sahara was green and dotted with lakes, like the one at the mouth of the "Cave of Swimmers" made famous by the book and film *The English Patient*.

Our own millennium started with the "Medieval Optimum," in which Vikings settled now icy Greenland and grapevines grew in Yorkshire, England. Then, from about 1550 to 1850, the so-called "Little Ice Age" took hold of Europe: Temperatures were around 1°C below those of the present century. The winter landscapes of Pieter Brueghel, depicting the Dutch countryside frozen over and blanketed with deep snow were painted during this period. Lake Constance, Western Europe's largest lake, regularly froze over completely—something that has happened only once this century, in 1963. In England, festivals took place on the ice of the River Thames. The Great Frost of 1608 was later described in a novel by Virginia Woolf[2]: "Birds froze in mid-air and fell like stones to the ground. At Norwich a young countrywoman started to cross the road in her usual robust health and was seen by the onlookers to turn visibly to powder and be blown in a puff of dust over the roofs as the icy blast struck her at the street corner. The mortality among sheep and cattle was enormous. It was no uncommon sight to come upon a whole herd of swine frozen immovable upon the road."

What role the ocean circulation played in these climate changes is as yet unclear. New data from sediment cores (Bond et al., 1997) strongly suggest that these events were part of a more or less regular 1,500-year-long cycle, involving major shifts in the North Atlantic ocean currents. With recent advances in computer modeling and the increase of data available on past climates, we may be on the verge of developing an understanding of such cycles.

THE FUTURE: RISK OF A SEA CHANGE?

The climate of the next century will be defined by an ongoing increase in the concentration of carbon dioxide and other greenhouse gases in the atmosphere. All climate models are predicting that this will lead to a substantial temperature increase (around 2°C by the year 2100) (Houghton et al., 1995). The hydrological cycle of evaporation and precipitation is also expected to increase, as in a warmer world the atmosphere can hold more moisture.

How will the Atlantic Ocean circulation respond to these changes? Given its past instability, this is a very real concern. A warmer climate will mean less cooling and more precipitation, possibly also extra freshwater from a melting of the Greenland ice sheet. The delicate balance in which the present conveyor operates may cease to exist. Model scenarios for the twenty-first century consistently predict a weakening of the conveyor by between 15 and 50 percent of its present strength (Rahmstorf, 1999).

What has not yet been determined is whether sudden temperature shifts like those seen in the Greenland ice core could take place. Due to their limited resolution, the current generation of climate models cannot properly represent the processes that might lead to these sudden changes (i.e., the convective instability). We do not yet know if or when we would cross a threshold in the climate system that could dramatically change our future (Rahmstorf, 1997). It is a risk we cannot afford to ignore.

NOTES

1. Strictly speaking, there is no net transport of salt between oceans in equilibrium but rather an exchange of salinity (salt concentration), which is a quantity routinely measured from ships.

2. Woolf's description in *Orlando* is based on a contemporary report by Thomas Dekker.

REFERENCES

Bond, G., W. Showers, M. Cheseby, R. Lotti, P. Almasi, P. deMenocal, P. Priore, H. Cullen, I. Hajdas, and G. Bonani. 1997. A pervasive millennial-scale cycle in North Atlantic Holocene and glacial climates. *Science* 278:1257–1266.

Broecker, W. 1987. The biggest chill. *Natural History Magazine* 97:74–82

Broecker, W. S. 1991. The great ocean conveyor. *Oceanography* 4:79–89.

Broecker, W. 1995. Chaotic climate. *Scientific American* 273:44–50.

Ellis, H. 1751. *Philosophical Transactions of the Royal Society of London* 47:211–214.

Ganopolski, A., S. Rahmstorf, V. Petoukhov, and M. Claussen. 1998. Simulation of modern and glacial climates with a coupled global model of intermediate complexity. *Nature* 391:350–356.

Houghton, J. T., L. G. Meira Filho, B. A. Caleander, N. Harris, A. Kattenberg, and K. Maskell. 1995. *Climate Change 1995—The Science of Climate Change.* Cambridge, U.K.: Cambridge University Press.

Labeyrie, L. D., J. C. Duplessy, J. Duprat, A. Juillet-Leclerc, J. Moyes, E. Michel, N. Kallel, and N. J. Shackleton. 1992. Changes in the vertical structure of the North Atlantic Ocean between glacial and modern times. *Quaternary Science Reviews* 11:401–413.

Manabe, S., and R. J. Stouffer. 1988. Two stable equilibria of a coupled ocean-atmosphere model. *Journal of Climate* 1:841–866.

Rahmstorf, S. 1994. Rapid climate transitions in a coupled ocean-atmosphere model. *Nature* 372:82–85.

Rahmstorf, S. 1995. Bifurcations of the Atlantic thermohaline circulation in response to changes in the hydrological cycle. *Nature* 378:145–149.

Rahmstorf, S. 1996. On the freshwater forcing and transport of the Atlantic thermohaline circulation. *Climate Dynamics* 12:799–811.

Rahmstorf, S. 1997. Risk of sea-change in the Atlantic. *Nature* 388:825–826.

Rahmstorf, S. 1999. Shifting seas in the greenhouse? *Nature* 399:523–524.

Stommel, H. 1961. Thermohaline convection with two stable regimes of flow. *Tellus* 13:224–230.

Thompson, B. 1798. Pp. 237–400 in *The Complete Works of Count Rumford*. 1870 reprint ed., Boston, Mass.: American Academy of Sciences.

Warren, B.A., and C. Wunsch. 1981. Deep circulation of the world ocean. Pp. 6–40 in B. A. Warren, ed., *Evolution of Physical Oceanography*. Cambridge, Mass.: MIT Press.

Weaver, A. 1995. Driving the ocean conveyor. *Nature* 378:135–136.

14

Simplicity and Complexity

Murray Gell-Mann

I congratulate Mercedes Pascual and Stefan Rahmstorf on their winning the competition in the area of Global and Complex Systems, on their prospects for great achievements in the future, and on their excellent contributions to this volume. They were chosen for the McDonnell Centennial Fellowships because of the impressiveness of their application essays, the importance of their topics, their promise as individuals, and the difference the award would make to them.

The Foundation insisted that the application essays themselves play a very important part in the selection process. But the competition was international, and therefore special allowance was supposed to be made for those applicants who did not have English as a native language. It turned out, however, that many of the best-written essays in English were submitted by just such people, including the two recipients of these awards. Most of the native speakers of English who applied were Americans. Maybe American education at the primary and secondary level really does leave something to be desired.

Naturally, a number of other applications submitted in this competition were also quite impressive. As a member of the Advisory Committee, I wished that the Foundation had had greater resources so as to afford some additional expenditures. As usual in these cases, I felt it would have been nice to fund some more of the best applicants. But in addition, and that is perhaps not quite so usual, I felt that it would have

Santa Fe Institute, Santa Fe, New Mexico

been wonderful to fund collaborations among various applicants. There were several cases in which the researchers who wrote to us were tackling complementary aspects of the same situations. For example, some of them were concerned with different parts of the huge field encompassing global climate and the behavior of the oceans. Others treated various aspects of ecology, including fires of natural and human origin, changes in the character of lakes, and so on. What an appealing idea to bring these various scientists together to supplement their ongoing individual research projects with collaborative work exploiting the strong links among their topics of study. Even the two winners might benefit from exploring the connections between their projects.

Our Advisory Committee spent a good deal of time choosing the phrase "Complex and Global Systems" and discussing its meaning. In addition, unlike the other advisory committees, we met for an extra day and held a seminar on our topic.

By global systems, we mean, of course, those that relate to the biosphere of our planet as a whole and to the life forms in it, including us human beings. Both of the chapters in this section have to do with global systems.

The word "complex" is a little trickier to interpret. Some people use the term "complex systems" to refer to anything composed of many parts (even if those parts are rather similar, like ions in a crystal). I believe that is an abuse of language unless the parts exhibit a great deal of diversity.

In my view, and that of many of my colleagues, something is complex if its regularities necessarily take a long time to describe. It then has what I call "effective complexity." The plot of a novel is complex if there are many important characters, a number of subplots, frequent changes of scene, and so forth. A huge multinational corporation is complex if it has many subsidiaries selling different goods and services, with a variety of management styles suited to the various countries in which they operate. The United States tax code is complex. So is Japanese culture.

Note that effective complexity refers to the minimum description length for the regularities, not the random or incidental features. Randomness is not what is usually meant by complexity.

In my talks on simplicity and complexity, I often make use of three neckties to illustrate my ideas. One of those ties has a very simple pattern of stripes in three colors, repeated over and over again with the same spacings between the stripes. The next one, (an Ermenegildo Zegna tie, the same brand Monica is supposed to have given the President) has a more complex pattern—its regularities would take somewhat longer to describe. The last one, a hand-painted tie from Austin, Texas, has a very complex pattern, with its regularities requiring a very long description indeed. In this discussion I restrict myself to the pattern, ignoring things like soup stains, wine stains, and so forth.

But the choice of features to study is to some extent subjective. To a dry cleaner, for example, the pattern of the tie may be incidental, while the character of the stains is a crucial regularity.

How does one separate the total information about something into a part describing regularities and a part describing random or incidental features? Much of the discussion in the chapters by Pascual and Rahmstorf is precisely about this sort of issue. Where is a practical place to draw the line between pattern or regularity and what is treated as random or incidental?

A very useful way to characterize the regularities of an entity is to imagine that it is embedded in a set of entities that display its regularities and differ in incidental or random features, the entity being described and the other, imaginary ones are assigned probabilities. The set is then what is called an "ensemble," as, for example, in statistical mechanics. Thus we arrive at a two-part description of the entity: first the ensemble to which it belongs and then the address—within that ensemble—of the specific entity itself. That is similar to what we do on a computer when we utilize a basic program and then feed specific data into that program. The regularities are like the basic program and the incidental features are like the additional data.

Those of us who study the fundamental laws that govern the behavior of all matter in the universe seem to be finding that those laws are very simple. However, they are probabilistic rather than fully deterministic. The history of the universe is thus codetermined by those simple laws and an inconceivably long string of chance events—"accidents" that could turn out in different ways with various probabilities. In this way the history we experience is embedded in an ensemble, a branching tree of possible histories, with the accidents at the branchings. In a splendid short story by Jorge Luis Borges, someone has constructed a model of the branching histories of the universe in the form of a Garden of Forking Paths.

In the history we experience, when each chance event happens, only one of the possible outcomes is selected. Before it happens, then, each future accident is a source of unpredictability, of randomness. But once it occurs, a past accident may be an important source of regularity.

Some of those chance events produce a great deal of regularity, at least over a considerable stretch of space and time. I call those events "frozen accidents." They are responsible for most of the effective complexity that we see around us.

Think of all the accidents that have produced the people who have contributed to this volume. Start with the little quantum fluctuation that led to the formation of our galaxy. That fluctuation was presumably not very important on a cosmic scale, but it was very important indeed to anything in our galaxy.

Then there were the accidents that led to the formation of the solar system, to the structure of our own planet, to the origin of life here on Earth, to the particular course of biological evolution, to the characteristics of the human race. Consider also the accidents of sexual selection, of sperm meets egg, that led to the genomes of everyone here. And then the accidents of development, in the womb and in childhood, that produced the adult human beings.

An example from biology of what seems to be a frozen accident is the fact that right-handed sugars and left-handed amino acids play important roles while the corresponding mirror image molecules do not. Some theorists have tried to account for that asymmetry by utilizing the left-handed character of the weak interaction for matter as opposed to antimatter. However, in forty years no one has succeeded, as far as I know, in making such an explanation stick. It seems that the asymmetry must be an accident inherited from very early forms of life.

Among those few historians who are willing to consider contingent history (speculating about what would have happened if something had gone differently), it is fashionable to discuss an incident that occurred in 1889. Buffalo Bill's Wild West Show was touring Europe and reached Berlin. One of the most important acts was, of course, that of Annie Oakley, the female sharpshooter. She would ask for a male volunteer from the audience who would light up a cigar that she would then shoot out of his mouth. Generally, there were no such volunteers and her husband, himself a famous marksman, would step forward instead. On this occasion, however, there was a volunteer, the Kaiser, Wilhelm II, who had succeeded to the throne the previous year upon the premature death of his father. He took out an expensive Havana cigar, removed the band, clipped off the end, and lit it, waiting for Annie to shoot. She was a bit worried, having drunk heavily the night before, but she took aim and fired. We know the result. But what if things had gone differently?

Work on complex systems involves a mix of general principles and of specific details that result from history, indeed from historical accident. Take the study of the oceans and global climate. The rotation of the Earth matters, along with its period—the day—the length of which has varied over time. The length of the year and the inclination of the equator relative to the plane of the Earth's orbit also matter. So do the layouts of the oceans and the continents. (Those, too, were different in the past, of course.) There are specific ocean currents and specific wind systems, and so on and so forth. That is what makes for complexity. In comparison, the basic physics of the air and the water is in general fairly simple.

However, another consideration enters besides effective complexity. How hard is it to calculate the consequences of the physical laws? How much computation is necessary to solve the equations to the needed accuracy? That is related to what is called logical depth.

Often it is not easy to tell whether we are dealing with effective complexity or with logical depth. When I was a graduate student fifty years ago we wondered what kind of dynamics underlay the structure of atomic nuclei. It appeared that the laws governing the energy levels of nuclei would turn out to be extremely complex. Today, however, we have the theory of quantum chromodynamics—the field theory of quarks and gluons—supplemented by quantum electrodynamics, which describes the electromagnetic interaction. We now believe that the combination of those simple theories yields the energy levels of nuclei to an excellent approximation, but the calculations are so elaborate that even today's machines and techniques cannot handle them properly. If, as we believe, the theories are correct (and many of their predictions have been verified by experiment), the energy levels of nuclei have very little effective complexity, but a great deal of logical depth.

Just as apparent complexity may reflect either effective complexity or logical depth or both, so too can apparent randomness reflect either fundamental indeterminacy or an effect of ignorance. Even in the classical deterministic approximation, indeterminacy can arise as a consequence of coarse graining, which refers to the level of detail at which a system is being studied. Neither observation nor calculation can be carried out with perfect accuracy, and so there is necessarily some coarse graining. But the coarse-grained system is no longer deterministic, because the coarse-grained past does not determine the coarse-grained future, but only the probabilities of various coarse-grained futures. That effect is particularly striking when the system is nonlinear and exhibits the phenomenon known as chaos, in which variations in the tiniest details of the initial state can produce big changes later on.

In connection with chaos, I am reminded of an incident that took place on the European tour promoting my book, *The Quark and the Jaguar*. I stopped in Barcelona, where my Spanish editor Jorge Wagensberg was also the director of the Science Museum, which had a wonderful exhibit on chaos. It featured a nonlinear version of a pendulum. The visitor was supposed to grab hold of the bob, let it go from a certain position and with a certain velocity, and watch the motion, which was also recorded on a drum by a pen. Then the visitor was supposed to grab the bob again and try to repeat the operation, reproducing as closely as possible the original position and velocity. No matter how carefully that was done, the subsequent motion came out quite different from what it was the first time, beautifully illustrating the meaning of chaos in science. While I was playing with the nonlinear pendulum, I noticed two men who were idling nearby, and I asked Jorge what they were doing there. "Oh," he replied, "those are two Dutchmen waiting to take away the chaos". It turned out that the exhibit was to be taken to Amsterdam as soon as I finished inspecting it. But I have often wondered whether many wealthy organizations wouldn't pay huge fees to these Dutchmen who could take away chaos.

The coupled systems of oceans and climate discussed by Stefan Rahmstorf and the coupled systems treated by Mercedes Pascual clearly exhibit both effective complexity and logical depth. Even though the basic equations involved are not that complicated, the specific features of the planet contribute a great deal of effective complexity, and the calculations are very difficult. As a result, a great deal of ingenuity is required in setting up the approximations, especially in choosing the right level of coarse graining in time and space and other variables.

Essentially, one is constructing an approximate model. Indeed, we have learned from our winners a great deal about the fine art of modeling, in connection with the dynamics of cholera and of fisheries—both interacting with the physical and chemical properties of bodies of water—and in connection with the interplay of climate, vegetation, and ocean currents.

At the Santa Fe Institute, which I helped to start and where I now work, researchers from a great many disciplines come together to investigate theoretical problems connected with simplicity and complexity and, in many cases, with complex systems. They construct models of real complex systems. (Usually these are computer models, but sometimes they are analytical ones or a mixture of the two.) Often these models are highly simplified. But how does one compare simple approximate models of complex systems with observation? What can one claim on the basis of such a model? Because it is very approximate, we would be embarrassed to have it agree accurately with observation! At our institute, we have had many discussions about these questions.

One answer is that it is often possible to identify features of the real system that obey, to a reasonable degree of accuracy, phenomenological laws that can be traced all the way from that real system through a sequence of approximations to the simple model under consideration. Then one can claim for the model that it helps to explain at least those features and those phenomenological laws. (Scaling laws are an example.) In some cases, one can also get a feeling for the level of accuracy to be expected for predictions based on the model.

When complicating a model by adding details, parameters, or assumptions, one has to decide, of course, whether the additional predictability is worth the extra complication. At the ends of the spectrum of detail, absurd situations are encountered. Mercedes Pascual, in that beautiful quotation from Borges, refers to one of those ridiculous limits, in which the model is simply a faithful reproduction of the entire system. At the other end of the spectrum one finds models that are so vague and general that they refer, very approximately, to a generic planet and lack most of the real characteristics and history of our own Earth. It is necessary to strike a wise compromise and avoid approaching either limit.

Both chapters contain just this kind of reasoning. Incorporating descriptive scaling laws into dynamics, as Mercedes Pascual recommends, is a good example of what I am discussing here. Mercedes Pascual and Stefan Rahmstorf are facing up brilliantly to the challenge of finding suitable models, in ecology and in the system of climate and oceans, balancing considerations of complexity, accuracy, and computational difficulty. They are carefully weighing how much special information must be included and how much should be left out. This kind of work will constitute, in my opinion, one of the main foci of theoretical research in the early decades of the new century. The intricate dance of simplicity and complexity, regularity and randomness, defines one of the most important frontiers of science today, particularly when it is applied to the history and likely future of our planet and its life forms, including us human beings.

Index